WHAT IS BPM?

WHAT IS BPM?

MARVIN M. WURTZEL

New York   Chicago   San Francisco   Lisbon   London   Madrid
Mexico City   Milan   New Delhi   San Juan   Seoul
Singapore   Sydney   Toronto

Library of Congress Cataloging-in-Publication Data
Wurtzel, Marvin.
   What is BPM? / Marvin Wurtzel.
      p. cm.
   ISBN 978-0-07-180225-3 (alk. paper)
      1. Reengineering (Management)   2. Organizational effectiveness.   3. Workflow—
Management.   4. Industrial management.   I. Title.
   HD58.87.W97 2013
   658.5'—dc23   2012039188

McGraw-Hill books are available at special quantity discounts to use as premiums and sales promotions, or for use in corporate training programs. To contact a representative please e-mail us at bulksales@mcgraw-hill.com.

What Is BPM?

1 2 3 4 5 6 7 8 9 0 QFR/QFR 1 9 8 7 6 5 4 3 2

ISBN 978-0-07-180225-3
MHID 0-07-180225-8

The pages within this book were printed on acid-free paper.

Sponsoring Editor
Judy Bass

Acquisitions Coordinator
Bridget Thoreson

Editorial Supervisor
David E. Fogarty

Project Manager
Virginia Carroll,
North Market Street Graphics

Copy Editor
Virginia Carroll,
North Market Street Graphics

Proofreaders
Virginia Landis
Sue Miller
Stewart Smith

Production Supervisor
Pamela A. Pelton

Composition
North Market Street Graphics

Art Director, Cover
Jeff Weeks

ABOUT THE AUTHOR

Marvin M. Wurtzel is a principal consultant for Marvin M. Wurtzel & Associates, a business process management consulting firm. He is a process improvement expert with many years of experience and has consulted, trained, and facilitated process improvement teams for such firms as ATT, Merck, JPMorgan Chase, Fidelity Investment, and the U.S. Coast Guard. Mr. Wurtzel is a Fellow in the American Society for Quality, a Malcolm Baldrige National Quality Award Examiner, a Certified Quality Engineer, a Certified Reliability Engineer, and a Master Black Belt (Six Sigma). He is a regular speaker and instructor at the BPM Institute and is currently an adjunct member of the faculty at Florida Atlantic University, in Boca Raton, Florida. He was previously an instructor for National Graduate School, in Falmouth, Massachusetts, and Northeastern University, in Boston, Massachusetts.

CONTENTS

INTRODUCTION TO BUSINESS PROCESS MANAGEMENT

INTRODUCTION

In many organizations, businesses, nonprofits, and government agencies, the need for higher productivity, better quality, and speed has caused managers to look for techniques and methods to make improvements. This has prompted organizations to adopt methodologies such as Lean principles, Six Sigma, Lean Six Sigma, and Total Quality Management (TQM). While these techniques have led to significant improvements in some organizations, many others have been unable to sustain the momentum. The underlying problem relates to the disconnect between the execution of their strategy and the day-to-day processes that produce value for customers and returns to stakeholders. The primary reason for this disconnect is that these organizations fail to look at their interconnected processes as a whole. They only look at the software development process, or the business transaction process, or some other specific function of the business. It is important to improve the entire gamut of business processes to achieve the desired competitive edge.

The strategy of an organization provides it with *what* it should be doing. The processes of the organization provide it with *how* it should be doing that, or the tactical day-to-day operation of the organization. The solution to the problem of improvement can be found in creating a framework for the business processes. This framework must emphasize the capture and documentation of the business processes and the metrics defining the required level of performance for each of those business processes. In other words, the organization needs to focus on improving the maturity level of key business processes.

LINKAGE TO ORGANIZATIONAL PERFORMANCE

Most management teams believe that their organizations are process oriented, and they understand how their products or services are delivered to their customers. The reality is that they don't really understand how the work is actually performed. Business Process Management (BPM) is a methodology that provides the management team and the organization as whole with a clear perspective on how the work is actually performed.

Organizational performance requires managers to understand and execute strategy. The inability to execute strategy can be blamed on many factors:

- The economy
- The competition
- The traditional mind-set of the organization
- The marketplace

And so on. When this happens, managers embark upon all sorts of efforts, which are detailed in the current literature, to fix the problem. They might institute an initiative such as Six Sigma, Lean principles, or process reengineering, but if they don't have a clear picture of how their products or services are produced, they will get only marginal results. BPM provides management with a clear picture of how the work is performed.

WHERE BPM FITS

Business Process Management (BPM) consists of defining and managing the end-to-end, or value chain, processes of the orga-

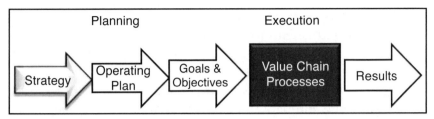

FIGURE 1-1. The flow from strategy to results.

nization in order to achieve the outcome of improved execution. While this sounds very straightforward, the reality is that it either is not done or is poorly executed in most organizations. Figure 1-1 shows the flow of strategy to operational plans and indicates that the value stream processes are a gray area, but, when executed properly, they can provide the organization with the results required. BPM sounds simple, but it conflicts with traditional management thinking. The vast majority of management teams think of their organizations in terms of the functional organization chart. They have little or no understanding of how the work is actually accomplished.

Many organizations look at BPM as a means to automate their processes because they have made extensive investments in their software systems and have approached processes with automation tools that route data from one system to another, such as automated interfaces between enterprise resource planning (ERP), customer relationship management (CRM), and supply chain management (SCM) applications. However, executing a project to document business processes often requires more than just tying together these systems to create a "lights-out" automated process.

In reality, not all complex business processes can be fully automated, for the following reasons:

• People are an integral part of the process.
• Mistakes and exceptions occur unpredictably within the process.
• Complex process steps are not easily reduced to digital business rules.

Any business process initiative needs to start with a strong process management component, such as BPM, to identify which processes require improvement, redesign, or reengineering to eliminate, or reduce, the steps that cause the greatest number of errors, result in wasted time or resources, or represent the most cost to the company. The processes may be automated, but as mentioned, the answer to process improvement is not always simply process automation. While it does incorporate process automation, this is not the sole purpose of BPM. Gaining control and managing your core business processes is the true objective of BPM.

The essential nature of any organization is people working together to achieve a common goal. For the overall effort to be successful, decisions and actions must be coordinated among individual contributors and between functional organizations.

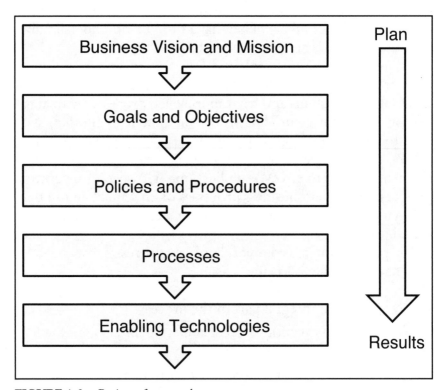

FIGURE 1-2.   Business framework.

They also must be consistent and yield satisfactory results at a reasonable cost. These factors are a result of the execution of business processes.

Generally, management teams follow an approach or work within a framework that is a top-down structure from planning to action, as illustrated in Figure 1-2, but it is typically focused on each functional organization, working in a silo. The real problem is that the processes that create value for customers are typically cross-functional in nature.

It should go without saying that the framework of most organizations also rests on an organizational commitment to quality. Six Sigma, Kaizen, Operations Excellence, Lean principles, Lean Six Sigma, and similar quality methodologies foster a continuous process improvement culture. The execution of these methodologies is typically performed in a functional organization setting. This tends to lead to suboptimizing processes that are, in reality, cross-functional in nature.

BPM FUNDAMENTALS

A simple definition of a business process is "a sequence of steps performed for a given purpose." Thus, a process is a predetermined course of action and a standing plan for people to follow in carrying out repetitive tasks in a systematic way. A process is a translation of general plans and policies into a standard pattern of decisions and actions. It establishes the actions required and the timing and sequence of activities. A business process can be described as a collection of related, structured activities or processes (a chain of events or activities) that produce a specific product or service for a particular customer or customers (internal or external to the organization). A business process usually cuts across several functions, for example, operations, logistics, sales, information technology (IT), finance, and legal.

BPM is a straightforward yet powerful method of looking beyond functional organizations and their activities and rediscovering the strategic or core business processes that actually make the business run—the value chain. It provides a discipline that enables us to find our way through the complexity of the organizational structure and focus on the business processes that are truly the heart of the business. Figure 1-3 illustrates a typical organization chart for a business with examples of value chain processes shown horizontally to indicate their cross-functional nature.

The reason for applying BPM is to understand how the organization creates value for its customers through a collection of processes that make up the value chain. These business processes are the day-to-day activities that produce the products or services for customers and ultimately generate income and, ideally, profits. The management, improvement, and application of technology are the keys to success for any organization.

BPM provides a graphical representation of what needs to be managed, with an appreciation for the sequence of activities and a "right" level of detail. What is documented and measured is what

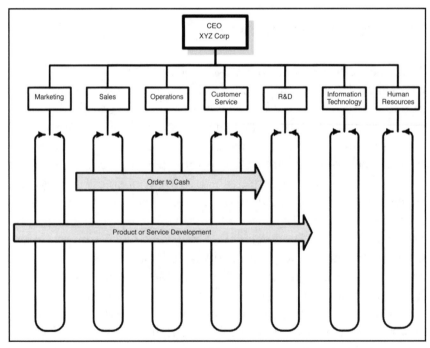

FIGURE 1-3. A typical organization chart showing examples of value chain processes.

gets done in any organization. By using the measurement information, organizations can make decisions based upon facts, which leads to better decision making and smarter problem solving.

The benefits of BPM are that it:

- Provides a framework to identify the core business processes, the value chain, which directly creates value for customers and returns to shareholders
- Builds a graphical representation of the key business processes, the linkages between them, and the dependencies that allow shared understanding within the organization
- Defines how the day-to-day activities (work) are performed across the organization to deliver value to customers
- Offers insight into the way key organizational functions perform business processes and the cross-functional interdependencies
- Offers insight into what needs to be measured and how to measure it, as well as a linkage to a reward system
- Allows for optimal application of enabling technology once all of the preceding tasks have been accomplished

An organization can use BPM in several ways. The first and most important is to understand how business processes interact in a system. Once it has accomplished this task, the organization can use the BPM methodology to locate issues that are creating systemic problems. This allows the organization to evaluate which activities add value for its customers and identify business processes that need to be improved or redesigned. Management teams can now use any of the various process improvement methodologies available—Six Sigma, Lean principles, process redesign, reengineering, and so on. The net result is to mobilize employee teams to streamline, improve, and manage business processes in an efficient manner and to successfully initiate significant change that improves business results.

What is different about BPM is that it breaks down the traditional view of the organization, represented by the functional bias of the organization chart. Most managers believe that the work or activities of the organization are executed at the micro level, within their control. The reality is that the work is done horizontally across the organization—that is, cross-functionally—with many organizations contributing to the end product or service. The other problem encountered with this view is an "inside-out" perspective of the organization, whereby department focus is the preoccupation. BPM builds a system model of the way an organization actually operates. This model shows all of the interactions between the core, or value chain, processes and highlights the cross-functional nature of these activities. Putting the appropriate measurements in place sheds light on how the organization is performing from the customers' viewpoint.

BPM provides the organization with a method to align the processes (the value chain) across the enterprise and optimize their performance. Applying BPM allows the organization to model the flow of data, people, systems, and physical resources and thus model the processes in alignment with the business objectives and market needs. BPM practitioners consider the value chain processes of the organization a strategic asset that must be understood, managed, and improved to provide value-added products or services to the customers.

CONCLUSION

BPM differentiates itself as a management approach by focusing on aligning an organization with its customers through the execution of processes. It promotes efficiency and effectiveness by providing an understanding of how the core processes interact and by striving for continuous improvement and integration with enabling technology. BPM can be described as enabling the organization to be more efficient and effective by becoming a process-based organization rather than by using the traditional functionally managed approach.

WHY DO WE NEED BPM?

THE IMPORTANCE OF BUSINESS PROCESSES

Business processes provide the link between functions in the organization. Processes play a prominent role in the coordination of activities between individual contributors and functional departments. Business processes operate horizontally and vertically along the lines of organization. Therefore, the output of one process becomes an input for another.

The business processes determine the costs of the products or services the organization is selling in the marketplace. It is cheaper to handle repetitive activities by using standard practices. Work can be explained and delegated to people easily and individual cases can be processed more quickly at a lower unit cost when processes are documented and maintained. There is less need for supervision, and administration becomes a simplified task.

Well-documented business processes ensure consistency of decisions and tasks of the same type when multiple actions take place. This is important for internal control and dealing with customers. Information used for making decisions and plans, and for taking guiding actions, must be reliable. Documented processes are needed for collecting, recording, storing, and transmitting information. If the processes are well designed, documented, measured, and maintained, the information will be reliable. These processes can provide adequate assurance to the top management that the activities that are to be done will be executed at the proper time in a proper way.

Many organizations and their senior management do not look at business processes as a critical resource and an asset. Business processes are a corporate resource, as are the physical plant and equipment, personnel, and so on, and they require management, measurement, and maintenance in similar fashion. Business Process Management can help provide those elements.

Most executives understand the organization as a collection of functions on an organization chart. The success that they have had, as well as their promotions, have come as the result of being good at their job in a given function—operations, marketing, sales, and so on. As these executives are promoted, they become removed from the routine activities that are performed as part of the normal business processes and they lose sight of the fact that work is accomplished through cross-functional processes. They show little or no understanding of how the business processes that accomplish work are actually organized.

Viewing a business with a functional bias has the following consequences:

- Increased
    o Cycle time
    o Inspection
    o Costs
    o Rework
    o Work-arounds
    o Exceptions
- Decreased
    o Efficiency
    o Quality
    o On-time delivery
    o Customer satisfaction

WHY DON'T MORE ORGANIZATIONS EXPRESS THEIR STRATEGY IN PROCESS TERMS?

The culprit for the failure of organizations to use process in formulating strategy is once again the traditional functional mind-set. This is manifested by leaders who are not accustomed to thinking in process terms at the enterprise level. Instead, they view processes at a micro level, more as procedures. Most managers think of process as something that provides systematic reproducibility, regardless of changes in personnel. It provides them with a sense of relief because there is nothing for them to worry about. They end up associating process with something that is cumbersome and rigid. It is the view of BPM practitioners that, in far too many organizations, senior management's traditional functional mind-set represents one of the most significant barriers to improving performance. Indeed, there is reason to believe that the traditional functional paradigm has done more to impede customer-focused business performance improvement over the past few decades than almost any other factor.

This mode of thinking prevents executives from understanding and improving the flow of cross-functional activities that create enduring value for customers and shareholders. It impedes the effective deployment of enabling information technology (IT). In fact, one could argue that it's the functional mind-set that has been at the core of many rather appalling business decisions in terms of deploying information technologies, as it reinforces thinking within functional silos, such that the information needs of downstream departments receive consideration only on a secondary basis—if at all. It also promotes turf protection and an undue preoccupation with organization structure.

This traditional mind-set contributes to the belief that it is somehow possible to improve an organization's performance by properly defining the boxes on the organization chart and filling in the names of the "right" people in the key boxes. (See Figure 2-1.) Little could be further from the truth. This thinking encourages a distorted view of performance measurement

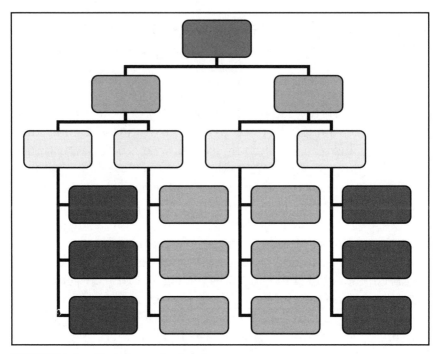

FIGURE 2-1. Typical organization chart.

and executive rewards, shifting focus away from meaningful measures such as the timeliness and quality of services provided to customers and toward less significant measures around functional or departmental performance.

BPM is a simple yet powerful method of looking beyond functional activities and rediscovering the strategic or core business processes that actually make up the value chain and allow managers to run the business. It is a discipline that enables them to peel away the complexity of the organizational structure and focus on the business processes that are truly the heart of the business. (See Figure 2-2.)

An organization—any organization—creates value for customers through a collection of business processes. These business processes are the natural business activities that produce value by providing products or services, serve the customers, and, if executed effectively, generate income and returns to the shareholders. Managing, controlling, and improving these business processes are the keys to the success of any organization.

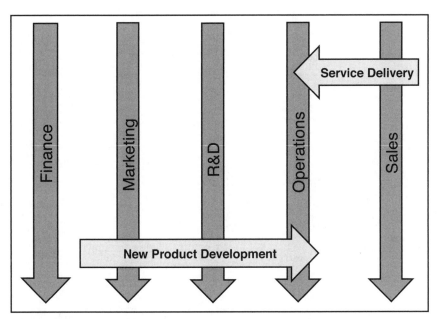

FIGURE 2-2. How work is done.

HOW IS BPM DIFFERENT?

BPM differs from other management methodologies in its mental model of what needs to be managed—that is, the understanding of how these processes interact in the system. BPM forces the management team to have a much greater appreciation for the sequence of the activities and the appropriate level of detail. Once critical processes are documented, managers can create measurements of performance—key process indicators (KPIs)—based on the timeliness, quality, and cost of products or services provided. The nature of decisions and the information on which decisions are based lead to a different set of dynamics in problem solving and decision making.

BPM is based upon the concept that the processes of an organization are a strategic asset that needs to be managed and supported. This support comes from the fact that the processes must be customer centric, business driven, and data driven, and the foundation for them is continuous improvement. (See Figure 2-3.)

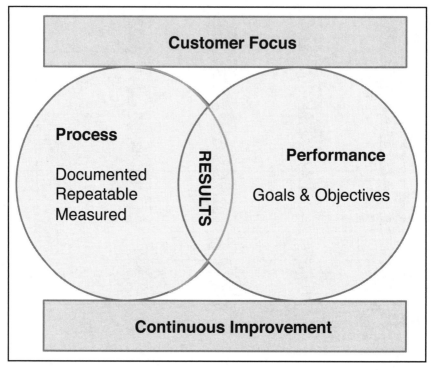

FIGURE 2-3.   The Business Process Management model.

Customer centric. Customers are one of the biggest stakeholders in the process output. The processes should be targeted to meet or exceed the customers' requirements.

Business driven. The processes are meant to meet the requirements of the business strategy. Their capability of meeting required performance levels is a determining factor in profitability. The output of these processes is expected to be timely and defect- or error-free. The processes need to be managed in terms of volume, accuracy, cycle time, efficiency, quality, and so on.

Data driven. Management by fact is necessary in today's business world. A data-driven approach enhances managers' decision-making capability, which in turn enables them to make quick and timely decisions.

Continuous improvement. Improvement is essential in business. Continuous improvement is the minimum requirement in a competitive environment with demanding customers. As globalization continues, profit margins are squeezed, directly impacting the bottom-line results for the organization. An approach based on continuous improvement looks for opportunities to constantly improve processes and extract the best value for customers.

As you gain an understanding of the value chain processes, you need to ensure that they are focused on satisfying the customer—in other words, that they are customer centric. They must also support the business drivers and be measured in such a way that the data drives the organization to make the necessary improvements or changes.

WHAT DOES BPM PROVIDE?

BPM provides a shared understanding of the key business processes, their linkages, and their dependencies. It enables management to understand how work is performed across the organization to deliver value to customers and determine the correct skill sets required for employees, the appropriate technology tools to implement, and the appropriate business rules required for repeatable process execution.

BPM provides the organization with a framework by which to identify the core business processes that directly create value for customers and returns to shareholders—that is, the value chain. Management can develop insight into key organizational and business process interdependencies. BPM provides a clear picture of the cross-functional nature of the work activities, enabling managers to clearly understand what needs to be measured, which can then be linked to the reward system of the organization. Last, BPM allows for the optimization of enabling technologies.

HOW DO YOU USE BPM?

You can use BPM to understand how business processes interact in a system and to locate issues that are creating systemic problems and reducing your ability to operate efficiently, which in turn reduces your ability to deliver on time or results in poor-quality products or services. You can apply BPM to:

- Evaluate which activities add value for customers
- Identify business processes that need to be redesigned
- Mobilize teams to streamline, improve, and manage business processes

The bottom line is that BPM can successfully initiate significant change that improves business results. Value chain or core processes need to be aligned with an organization's business strategies. This alignment results in a push for significant improvements in efficiency and effectiveness. Integrating strategy, processes, metrics, improvement projects, and IT into a shared objective of driving improved bottom-line results yields tremendous value. BPM helps to provide the view of the organization that puts all of the pieces into perspective.

Processes are fundamental to any organization. The capabilities and performance of these processes are important parameters in defining the success or failure of the enterprise. BPM offers management at all levels of the organization the following benefits:

- A view of the work performed from input to output
- Visible control of the processes
- An early warning system to detect developing problems
- Enhanced ability to make day-to-day decisions

SUMMING IT ALL UP

BPM relies on the concept of defining the macroprocesses, assigning ownership, and creating responsibilities for the owners, monitoring, changing, or redesigning processes to deliver greater value or optimal results, through continuous improvement. BPM facilitates strategically aligned, bottom-up measures that can be used to provide useful information to management on performance against business goals.

BPM, in broad terms, facilitates:

- Understanding of the processes and the process owners' responsibilities
- Measuring and monitoring of the processes in relation to their strategic alignment
- Mitigation of risk by providing feedback on performance
- Fast and data-driven decision making by management

UNDERSTANDING PROCESSES

THE BASICS OF PROCESS

The value chain of an organization operating in a specific area of business or government is the collection of processes required by the organization to provide products or services to its customers. The processes are collections of related activities performed by employees of the organization for the purpose of achieving a common business goal: providing the product or service to their customers. Products or services to be provided to the customers must pass through all the activities of the chain in a specific order, and at each activity the product or service gains some value. The chain of activities gives the products more added value than the sum of the added values of all the individual activities. As a rule of thumb, the business unit—not the divisional level or corporate level—is the appropriate level for construction of a value chain.

The experts in this field, such as Thomas H. Davenport, author of *Process Innovation: Reengineering Work through Information Technology* (Harvard Business Review Press, 1992), defines a (business) process as "a structured, measured set of activities designed to produce a specific output for a particular customer or market. It implies a strong emphasis on how work is done within an organization, in contrast to a product focus's emphasis on what. A process is thus a specific ordering of work activities across time and space, with a beginning and an end, and clearly defined inputs and outputs: a structure for action. . . . Taking a process approach implies adopting the customer's point of view. Processes are the structure by which an organization does what is necessary to produce value for its customers."

• Account Management	• Customer Life Cycle Mgmt.	• Materials Storage	• Regulatory Document Mgmt.
• Adv. Planning & Scheduling	• Customer Requirements ID	• Order Dispatch & Fulfillment	• Repair Planning Management
• Advertising	• Customer Self-Service	• Order Management	• RFQ & Contract Management
• Assembly	• Customer/Product Profitability	• Organizational Learning	• Returns Management
• Asset Management	• Demand Planning	• Payroll	• Sales Channel Management
• Benefits Administration	• Distribution/VAR Management	• Performance Monitoring	• Sales Commission Planning
• Branch Operations	• Expense Report Processing	• Performance Review	• Sales Cycle Management
• Budget Control	• Financial Planning	• Physical Inventory	• Sales Planning
• Build to Order	• Financial Close/Consolidation	• Planning & Scheduling	• Service Agreement Management
• Call Center Service	• First Day/Last Day Mgmt.	• Postsales Service	• Service Fulfillment
• Capacity Reservation	• Hiring/Orientation	• Problem/Resolution Mgmt.	• Service Provisioning
• Capital Expenditures	• Installation Management	• Process Design	• Shipping Logistics
• Claims Processing	• Integrated Logistics	• Procurement	• Site Survey & Solution Design
• Claims Liquidation Processing	• Internal Audit	• Product Data Management	• Sourcing
• Check Request Processing	• Inventory Management	• Product Design & Dev.	• Station Repair Management
• Collateral Fulfillment	• Investor Relations	• Product Documentation Mgmt.	• Strategy Development
• Commissions Processing	• Invoicing	• Product Launch Management	• Succession Planning
• Compensation	• IT Service Management	• Product Life Cycle Mgmt.	• Supply Chain Planning
• Component Fabrication	• Knowledge Management	• Product/Brand Marketing	• Supplier Relations Management
• Corporate Communication	• Manufacturing	• Professional Service Mgmt.	• Timekeeping/Reporting
• Credit Request/Authorization	• Manufacturing Capability Dev.	• Program Management	• Training
• Customer Acquisition	• Materials Expense Mgmt.	• Property Tracking/Accounting	• Treasury/Cash Management
• Customer Account Mgmt.	• Market Research & Analysis	• Publicity Management	• Warehousing
• Customer Help Desk	• Market Test	• Real Estate Management	• Warranty Management
• Customer Inquiry	• Materials Procurement	• Recruitment	• Zero-Based Budgeting

FIGURE 3-1. Processes in an organization.

While this sounds simple, organizations generally manage these activities with great difficulty, resulting in poor execution. The primary cause for this is the lack of understanding of how certain processes are relevant in an organization and exactly what needs to be managed. If you ask the management of a given organization what processes exist there, the list would resemble that shown in Figure 3-1, which contains approximately 100 entries. To manage the execution of those processes would require an enormous effort.

THE NUMBER OF BUSINESS PROCESSES IN A TYPICAL COMPANY

Most Business Process Management practitioners will tell you that there are between five and ten strategic or core processes in any organization. Michael Hammer and James Champy once observed that "hardly any company contains more than ten or so principal processes." They were referring to the high-level strategic processes, which are just the tip of the iceberg in any large business. Under the surface exist hundreds of tangible processes that actually run the business.

Beyond that limited number of strategic processes there are the support and management processes that exist in the organization. Those processes are needed to run the organization but do not provide any direct value to the customer.

TYPES OF PROCESSES

Even though most organizations have not done a good job of documenting their business processes, there is a considerable amount of employee knowledge embedded in a given company's processes. One method of understanding this knowledge involves classifying processes based on a framework. One such framework classifies processes into three categories: value chain, support, and management.

Value chain processes constitute the core business and create the primary value stream to the customer. Typical *support processes* support the value chain processes. *Management processes* govern the operation of the organization.

- Value chain, strategic, or core processes
  - o Operational
       Develop—make/buy—sell—deliver—service
- Support processes
  - o Enabling or resource management processes
       Financial—human—physical—material—information
          technology
- Management processes
  - o Control or governance processes
       Planning—budget—change—compliance

Support processes comprise those shown in Figure 3-2. The primary focus of BPM is on the value chain (interchangeably called *core* or *strategic*) processes of the organization.

• Human Resources
   – Payroll
   – Hiring
   – Merit Reviews
• Facilities
   – Maintenance
   – Cleaning
   – Power and Light
   – Safety
• Information and Technology
   – PC Support
   – Database Management
   – Technology Refresh
	• Finance
	   – Taxation
	   – Accounts Payable
	   – Accounts Receivable
	   – Stockholder Reporting
	• Procurement
	   – Equipment
	   – Materials
	   – Services

FIGURE 3-2. Examples of typical support processes.

DEFINING THE TERM *PROCESS*

Experts define the term *process* as follows:

—THOMAS DAVENPORT, 1992[1]
> *"In definitional terms, a process is simply a structured, measured set of activities designed to produce a specific output for a particular customer or market."*

—HAMMER AND CHAMPY, 1993[2]
> *"Process is a technical term with a precise definition: an organized group of related activities that together create a result of value to the customer."*

—PETER FINGAR, 2006[3]
> *"A business process is the complete and dynamically coordinated set of collaborative and transactional activities that deliver value to customers."*

---

[1] *Process Innovation: Reengineering Work Through Information Technology* (Harvard Business Review Press, 1992).
[2] *Reengineering the Corporation*, 1st ed. (HarperBusiness, 1993).
[3] *Business Process Management: The Third Wave* (Meghan Kiffer Press, 2006).

—WIKIPEDIA, 2012
> *"A Business process comprises a series or network of value-added activities, performed by their relevant roles or collaborators, to purposefully achieve the common business goal."*

Here is the definition of *process* that will be used for the purposes of this book:

> *A process is a collection of repeatable, value adding activities performed by human and technological resources of the organization for the purpose of achieving a common business goal, a product or service that the customer is willing and able to pay for.*

ORGANIZATIONS REPRESENTED IN THE VALUE CHAIN

The kinds of organizations typically represented in the value chain of a business are:

- Marketing
- Research and development
- Product or service design
- Operations
- Sales
- Distribution or logistics
- Service

A WORD TO THE WISE

When defining the value chain processes, avoid using the same terminology as the functional organization names. This helps those in the organization recognize that these processes are cross-functional and are not owned or tied directly to one organization.

As discussed, BPM is an enterprise-wide, structured approach to understanding how an organization provides the products or services to its customers. It is based on the premise that you must take a process view of your organization in order to understand what products and services your customer is willing to pay for. Based on that concept, BPM starts with defining the important cross-functional or strategic processes at an organization.

The process view (horizontal) versus the functional view (vertical) is shown in the preceding chapter in Figure 2-2.

VALUE CHAIN PROCESSES

The rationale for identifying processes that make up the value chain is very simple: they serve as a strategic weapon to achieve the organization's mission and vision by focusing on the critical business drivers. These processes frequently cut across organizational boundaries and involve numerous functional, departmental, or even divisional units. The understanding of their execution can serve to clarify inefficiencies in the organization. As a result of this understanding, the process is an end-to-end, cross-functional flow of activity.

Consider "order fulfillment" or "order to delivery" as a value chain business process. The subprocesses would then include order taking, scheduling, fulfillment, and shipping. An activity within order taking would be "enter order," and there would be several steps within "enter order" involving standards, instructions, forms, and skills, which would be executed in the sales organization. The activity of shipping would have its own set of standards, instructions, forms, and skills to be executed by the outbound logistics organization. As already stated, it takes several organizations to execute a value chain process, such as "order fulfillment" or "order to delivery," and BPM can be used to redeploy improvement efforts that were previously launched against relatively less important functional organization goals.

CHAPTER 4

BPM LIFE CYCLE PHASES

DEFINING THE BUSINESS PROCESS MANAGEMENT PHASES

The value chain processes are a strategic asset of the organization, so they must be documented, measured, improved (when necessary), and managed like any other asset. By following the methodology outlined, managers have the added advantage of using enabling technology to help execute these processes more efficiently and effectively.

Business Process Management is a systematic approach to understanding, improving, and managing an organization. It is generally accepted to have four phases: document, assess, improve, and manage. (See Figure 4-1.)

THE FOUR BASIC PHASES IN THE BPM DISCIPLINE

- *Document*   *process modeling*
- *Assess*   *analysis and measurement*
- *Improve*   *design/redesign*
- *Manage*   *control*

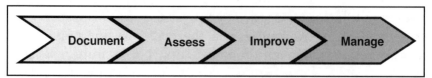

FIGURE 4-1. Business Process Management life cycle phases.

DOCUMENT PHASE

The *document phase* focuses on several things:

1. Identifying the value chain processes
2. Creating a process inventory
3. Classifying processes as either core, support, or management
   o Core processes are the value chain processes.
   o Support processes are those needed to run the business but that do not directly add value to the product or service.
   o Management processes are those required to supervise and monitor business operations.

It is essential in this phase to keep the view of the processes at a high level. This is what might be called the 100,000-foot view, as though you are looking out the window of an airplane and see large blocks on the ground below. This phase seeks to identify between 5 and 10 value chain processes and should not become overly detailed. The number of processes needs to be manageable—an organization cannot sensibly manage over 100 processes. Also, when you get to this level of detail, you are falling back into the functional mind-set, because at this lower level of detail the process may be defined in terms of the work performed within only one functional organization. At the ground-level view of the processes you can get tied up in too much detail—work instructions, policies and procedures, business rules, and so on. (See Figure 4-2.)

Now that the value chain has been identified, the next step is to create a system-level map for the organization. This system-level map is a high-level graphical representation of the sequence of work as it flows through the organization. The primary interest is in the core processes or the value chain of the organization. An example of a system-level process map is shown in Figure 4-2.

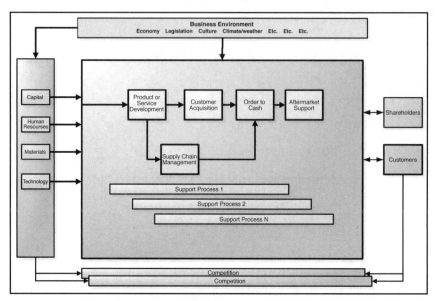

FIGURE 4-2.   System-level process map.

The system-level process map shown in the central box of Figure 4-2 identifies the value chain processes. Although this diagram seems to identify generic processes, it does identify the five core processes of an organization. The names assigned to these core processes should not be the same as those of any functions on the organization chart.

There are several ways to go about developing the system-level map. The first is to adopt models that are available in the existing literature in this field or that can be found through various research groups. An example of such a model is the MIT Business Activity Model. The other method is to create the system-level map by working with the senior executives of the organization and facilitating a discussion about how the organization creates value for its customers. The rationale for identifying the core processes is simple: they function as a strategic weapon to achieve the organization's mission and vision by focusing on the critical business drivers. Identifying and assessing them can serve to clarify any inefficiency in the organization structure. We can then deploy improvement efforts against the critical goals of the business.

In order to identify the core processes, certain attributes must be identified:

- Strategic importance
  - o Processes that have a major impact on the organization and are crucial to its overall success.
  - o Processes that are relevant to the strategy, mission, and goals of the organization.
  - o Processes that, when executed efficiently and effectively, can provide a competitive advantage and that, conversely, can be a disadvantage if not performed efficiently or effectively.
- Attributes
  - o *Customer impact.* Processes that directly affect customers are always considered operational core processes. They are the value chain. Typical examples are customer acquisition, order to cash, and aftersales support.
  - o *Cross-functional.* These processes frequently cut across organizational boundaries and involve numerous functional, departmental, or even divisional business units.

Once the team members reach consensus on the system-level process map, they can proceed to creating core process maps.

DOCUMENTING THE CORE OR VALUE CHAIN BUSINESS PROCESSES

The goal of this phase is to document, review, and allow for detailed analysis of the way a given business process is currently performed in order to determine whether improvement is required.

1. Define the purpose:
   - o What is the primary result of the execution of this process?

o   What product or service is provided to the customer of this process?

o   What constitutes an acceptable product or service?

2. Identify the boundaries:

o   What is the starting point of this process?

o   What is the trigger that starts this process?

o   What is the end point of this process?

o   How does the organization measure success?

3. Enlist the proper resources:

o   Who are the employees that directly participate in these processes?

o   Are these people who will speak freely about the actual performance?

o   Do these people understand the day-to-day details of the execution of the process?

4. Control details: Work at the 50,000-foot level, taking it to the next level of detail (medium level, 20 to 25 process steps). Don't get bogged down in excessive detail.

5. Revise for accuracy:

o   Review with other employees who did not participate in the original process documentation.

o   Review with customers of the process. They may be external customers—those procuring the product or service. They can also be internal customers—other employees who use the output in order to perform their functions in other processes.

The actual development of the core process maps requires that you have a clear understanding of the sequence of steps in the process. The best source of information to create this level of detail is the employees who work in this process on a day-to-day basis. You need representation of the key people/stakeholders in all the functional organizations that participate in this process, in order to provide the needed information. The preferred format for a process map is called a *swimlane diagram*—a flowchart, with lanes for each functional organization that participates. The process steps are mapped according to the organization performing the work. An example of a swimlane diagram is shown in Figure 4-3. Here again the boundaries of the process must be clear: where it starts and ends, who and what are the triggers that start the process, and who is the ultimate customer for this product or service. In developing the sequence of events, the handoffs between the subprocesses must also be identified. A clear definition item for information handed off at each step and the recipient of that information is essential to having a usable process map.

This step generally requires some facilitation by professionals who can keep the discussion and level of detail on track. Trained facilitators with little or no direct knowledge of the existing process are ideal because they ask more probing questions and do not have any preconceived knowledge of the information being provided. The most effective way to develop this type of process map is by hanging a large roll of paper on a wall and using Post-it Notes to develop the process flow. This allows changes to be made easily, while providing a document for all participants to work with. Once the team has come to a consensus on the

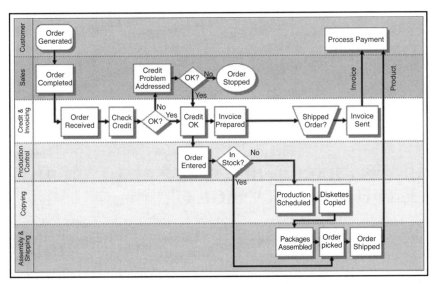

FIGURE 4-3.   Cross-functional process map/swimlane diagram.

process flow, the process can be documented in an appropriate electronic format.

After the core process has been documented, there are a few more things that need to be done. First is to validate the process that was documented, which is accomplished by walking through the process. This is a critical step because it ensures that what has been documented is executed. This is where any work-arounds, undocumented procedures, shortcuts, unnecessary activities, and so on, may be found. Once any necessary adjustments have been made to the process map, the configuration of this document is locked down and placed in a safe repository for future use.

The next task is to identify the process owners for the core processes. The owners must be at a level where they will have total accountability for the end-to-end execution of the process. The rule of thumb is that the process owner for a core process should be the individual with the most to gain or lose from its proper execution. The biggest pitfall in assigning ownership is taking a functional bias. This is why the core processes should not have the same names as functional organizations. An example is the sample core process map in Figure 4-3. That process could have been named "Sales," which would then fall under the functional bias criteria. Realistically, this process was executed by several organizational entities and was named "Order to Cash" for that reason. Ownership for this process could be assigned to someone from any of the organizations—Sales, Finance, or Operations—as shown on the swimlane diagram in Figure 4-3. Once the process owner has been identified, you are ready to proceed to the assess phase of BPM.

ASSESS PHASE—"IF YOU CAN'T MEASURE IT, YOU CAN'T MANAGE IT"

Once you have documented the value chain or core processes, you need to develop a set of measurements or key process indicators (KPIs) to determine whether the organization's performance can or does meet its strategy, goals, and objectives. There are three fundamental types of measurements that you can make for these processes: efficiency, effectiveness, and outcome.

Efficiency, also called *process measures*, are typically measured within the process. These measures represent parameters that directly control the integration of the resources. The resources may be human resources, equipment, consumables, space, electricity, parts, and so on. Process measures always include the performance of subprocesses. The most important thing to recognize here is that process measures enable you to predict characteristics of the output before they are delivered to the customers. These process measures enable you to make adjustments to the process in a way that prevents errors or defects persisting to the end of the process, where correction is most expensive.

Efficiency is multifaceted and is focused in different areas:

- Cost measures are intended to minimize the amount of resources consumed in the process.

- Variation measures are intended to eliminate the waste associated with non-value-added activities and insert contingency into plans and designs that cushion uncertainty.

- Cycle time measures are intended to reduce the total elapsed time required to transform input into outputs.

Process effectiveness, or *output measures*, quantify the capability of the process to deliver products or services according to their specifications and/or customer requirements.

Process effectiveness is determined by examining products or services after they are produced, so essentially you are looking in the rearview mirror. If you wait until the process is complete and all of the required resources have been applied and then find a problem, it is the most costly place to make the correction. The worst-case scenario would be if the problem escapes detection and is discovered by the customer. Not only do you have an expensive problem to correct, but you have also caused a customer satisfaction issue, which can be costly in the long run.

Process effectiveness measures quantify the likelihood of satisfying the customer's needs before the product or service is delivered. They are predictive in nature once you understand the customers' requirements and turn them into specifications.

Outcome, or *output (product or service) effectiveness measures,* determine how well the product or service performs for the customer by satisfying the needs and expectations of that customer. They are generally referred to as *measures of customer satisfaction.* Output effectiveness measures are retrospective and thus are obtained after the product or service is delivered to customers. They are necessary because they validate the process effectiveness measures that are being applied to control the day-to-day process activities.

Table 4-1 summarizes these three types of measures.

Now that you are acquainted with the types of measurements and the reasoning behind each type, you need to develop a set of process performance measurements, or key process indicators (KPIs). Here are the steps to follow:

1. Identify the output or outputs of the process.

2. Identify the customer base, and segment the customers when there are market or regional differences.

3. Learn the customers' requirements. Identify their needs and expectations by the various customer seg-

TABLE 4-1. Measuring Performance

If you can't measure it, you can't manage it!

TERMINOLOGY	EXPLANATION
1. Process or efficiency	1. Resources consumed in the process relative to minimum possible levels
2. Output or effectiveness	2. Ability of a process to deliver products or services according to specifications
3. Outcome or customer satisfaction	3. Ability of outputs to satisfy the needs of customers

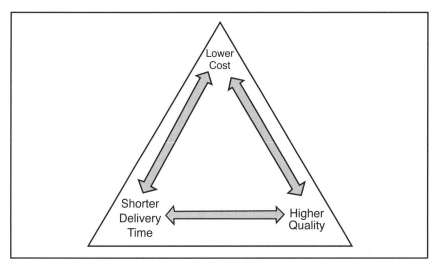

FIGURE 4-4. What customers desire.

ments. (See Figure 4-4.) Listen to the voice of the customers. Customers typically want products or services:

o   Faster = shorter time

o   Better = higher quality

o   Cheaper = less cost

Translate the needs and requirements into operational terms. Ideally, they must be things that can be measured or that are critical to quality characteristics. It is better to have quantitative measures because they can be translated into the process measures that represent parameters that directly control the integration of the resources. That enables you to predict characteristics of the output before it is delivered to the customers.

The characteristics to consider when measuring output quality vary, depending on whether the output is a product or a service. The following lists identify the characteristics that are critical to product and service quality, respectively.

Dimensions of Product Quality

1. Performance

2. Features

3. Reliability
4. Conformance
5. Durability
6. Serviceability
7. Aesthetics
8. Perceived quality or reputation of the organization

Determinants of Service Quality
1. Reliability
2. Responsiveness
3. Competence
4. Accessibility
5. Courtesy
6. Communication
7. Credibility
8. Security
9. Understanding of the customer
10. Tangibles

To reiterate: The output measures need to be quantitative. Product or service quality is characterized by its respective set of elements, shown in the preceding lists.

- Deliverables—attributes provided
- Interactions—customer experience

You need to ensure quality in both of these sets!

The objective of the *assess phase* is to determine measurements for the core processes. When an organization determines that there is an area where improvement is needed, management can direct improvement efforts to maximize benefit to the business. Process improvement is an organizational investment with a very nice payback, so you need to make the investment wisely. Chapter 5 contains a further discussion of process improvement.

In order to make decisions concerning process improvement or even automation, start with a quantitative assessment of the business's core processes. Then follow this procedure:

1. Create a gap analysis based upon these three criteria:
    1.1. Importance—what is the importance of making this improvement?
    1.2. Opportunity—how big is the opportunity if the improvement is successful (business case)?
    1.3. Feasibility—what is the likelihood of success? (This becomes critical in the initial stages of BPM implementation in order to create a model for further rollout.)
2. How to proceed:
    2.1. Rank the opportunities.
    2.2. Clarify the scope of improvement.
    2.3. Avoid suboptimization—don't improve one value chain process at the expense of another.
    2.4. Validate the choices.
    2.5. Differentiate product/service improvement from process improvement.
3. Develop and communicate an improvement/management plan.
    3.1. Create a shared understanding of beliefs, approach, and what needs to be done at the executive level and throughout the organization.
        3.1.1. Clearly state the vision.
        3.1.2. Clearly state beliefs, approach, and objectives.
        3.1.3. Communicate the business process framework and roles.
        3.1.4. Do it in multiple media.
        3.1.5. Do it frequently, clearly, and concisely.
        3.1.6. Create a sense of urgency.

3.1.7.   Focus on serving the customer.

3.1.8.   Lead by example.

3.1.9.   Don't overmanage.

IMPROVE PHASE

In a well-structured BPM initiative, the organization is driven by the need to efficiently and effectively satisfy the customers' needs and expectations. It is essential that the organization focus on customers and the processes that serve them—not on the CEO. This establishes process-oriented thinking, which is demonstrated in the organization's structure. You have identified process owners who are responsible and accountable for key processes. There is a drive to achieve excellence. The means and

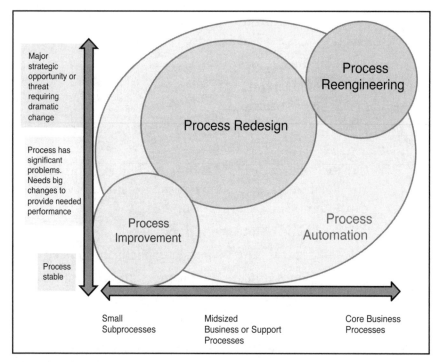

FIGURE 4-5.   Different levels of improvement projects.

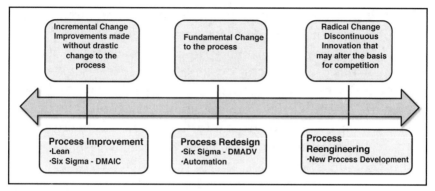

FIGURE 4-6. Process and level of change.

systems are in place to move in the stated direction. Goals and objectives are clear and meaningful.

In the *improve phase*, the goal is to evaluate how an organization's resources can be used most efficiently. Typical objectives are reducing costs, shortening cycle time, and improving product or service quality. Implementing a redesigned process can have adverse effects in other parts of the business, so care must be taken to make the improvements with an understanding of the entire value chain. Good use of analytical tools is required to execute the improvements correctly.

There are several choices when it comes to selecting the process improvement tools. Six Sigma, Lean principles, Lean Six Sigma, process simulation, process redesign, and process reengineering provide the necessary analytical tools if properly applied to the problems. (See Figures 4-5 and 4-6.) A good understanding of these tools is required for process improvement. Chapter 5 examines the fundamentals of these tools and the differences in their application.

MANAGE PHASE

The *manage phase* of BPM encompasses the tracking of the core processes, so that their performance can be easily monitored. The key process indicators are tracked and reviewed periodically to ensure that the processes are performing as required, and, if not, appropriate actions take place to make the necessary corrections. The primary responsibility to see that these things happen is in the hands of the process owners.

The KPIs are a combination of process effectiveness or outcomes, process efficiency, and customer satisfaction. The degree of monitoring depends on what information the business wants to evaluate and analyze. Decisions need to be made about whether you want to measure in real time, in near real time, or over some determined period. Some of these decisions are influenced by the level of automation applied to the processes.

The role of the process owners is absolutely critical at this stage. Their accountabilities can be divided into four areas: leadership, documentation, performance, and improvement.

Leadership

1. Drives the alignment between corporate strategy and customer focus
2. Prioritizes improvement opportunities
3. Resolves process interdependency issues
4. Owns process education and training
5. Leads the change to a process-focused organization

Documentation

1. Maintains documentation of the inputs and outputs of the process
2. Maintains the process documentation and approves all changes
3. Prioritizes automation
4. Ensures controls are in place for accurate financial reporting
5. Audits process compliance

Performance

1. Implements KPIs and reports process performance on a regular basis

2. Achieves KPI targets, goals, and objectives
3. Prioritizes performance gaps
4. Ensures adequate process resources
5. Monitors data collection system for accuracy

Improvement

1. Analyzes performance gaps
2. Develops plans to close the gaps
3. Identifies the appropriate process improvement tools based upon the gap
4. Executes process improvement projects, across the process
5. Benchmarks and adopts best practices
6. Fosters new process improvement ideas

The role of the process owners is critical to any successful effort to apply BPM. This is why it is important that the selection process for ownership be kept at a high level—the executive level—and should not revert to the functional mind-set.

CHAPTER 5

PROCESS IMPROVEMENT (IMPROVE PHASE TOOLS)

FUNDAMENTAL CONCEPTS OF THE IMPROVE PHASE

When it comes to process improvement, there are many choices in terms of how to approach it. Most organizations adopt a particular methodology and use it for all improvement projects. This is not always the best answer. Process improvement should be approached as toolbox from which to choose the right tool to do the job. This toolbox contains many different tools and requires different levels of skills, training, and understanding. These tools include process mapping and improvement, process reengineering, Six Sigma, and Lean principles. Each has its strengths and weaknesses, determined by the problem to be solved. This chapter provides an overview of each of these methods and points out the respective strengths and weaknesses.

Process improvement activities come with a price. That is, you must expend resources to make an improvement. These resources include human resources—that is, people who would otherwise be performing their day-to-day job functions, specialists trained in process improvement, or even external consultants brought in to train personnel and execute the problem resolution. This can be expensive, but the payback from a well-executed process improvement can save many times the cost of the improvement effort.

Let's look at an example that illustrates the source of the cost savings. Figure 5-1 shows the basic concept of profit and loss in a business. The first bar shows the simple concept for a business: there is a cost to produce a product or service, and then there is a profit margin on top. Reality is better represented by the bars to the right in Figure 5-1. The second bar shows the cost to produce the product or service, profit margin, and the associated waste. This is realistically where most organizations operate, and one of the goals of process improvement is to reduce the waste. The third bar represents the result of competition in the market-place. To be competitive a business must lower its selling price. If the business lowers its price without taking action on its cost to produce and its waste, then it also reduces its profit. If the business is forced into a significant price reduction, the result could be a loss instead of a profit.

Process improvement activities are intended to reduce waste and improve the processes used to create a business's products or services. If the business acts on those effectively by using the various process improvement tools discussed in this chapter, then it's possible to achieve a result that resembles the bar on the far right in Figure 5-1. Not only can the business reduce the waste

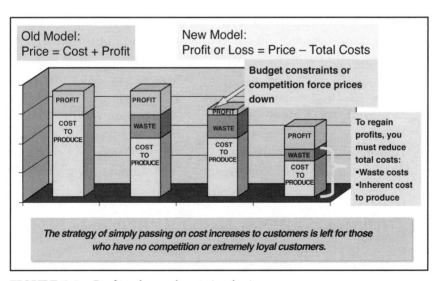

FIGURE 5-1. Profit or loss and waste in a business.

inherent in the process, it can also improve and simplify the process used to produce its products. This allows the business to maintain the profit margin and also potentially reduce prices and continue to be competitive in the marketplace.

Using process improvement tools in these areas creates opportunities for an organization to retain customers and provide them with what they want when they want it, as well as providing employees with the time and training to tackle work challenges.

Even a low percentage of defects or errors can result in numerous unhappy customers. Realistically, any organization wants to avoid any defects or errors that could result in the loss of revenue and poor customer satisfaction. Organizations might have been able to get away with high defect rates in the past, but it is definitely not a formula for long-term success. There has been a great deal of research suggesting that customers no longer sit around feeling sorry about their bad experiences with a product or service—they file a complaint and tell their friends and families about these negative transactions.

Keeping customers happy is good, and it is profitable for any organization. An increase in customer retention has been shown to increase profits by multiples of between 5 and 10. It has been shown that companies lose 15 to 20 percent of revenues each year to ineffective and inefficient processes—some suggest that this percentage might even be higher. Process improvement strives to make changes in order to meet short- and long-range organizational goals.

The earlier stages of defining and assessing the core processes of the organization operated at a relatively high level. At this point, the gaps in performance have been identified and the need to improve has been determined. It is time to start on the segments of those core processes that require improvement, which means moving to the "ground-level" view of the processes.

PROCESS MAPPING AND IMPROVEMENT

The goal of the methodology is to understand the current state, or "as-is," process, with the aim of documenting, reviewing, and analyzing the way a given business process is currently performed in order to improve upon it. The *current state process analysis* occurs in a team environment, using subject matter experts (SMEs) who work in this process. This serves as the basis for *future state process design* and *business case development.* When there are many critical areas that need attention, it is useful to develop a current state process map.

The current state process mapping session usually requires a facilitator (someone knowledgeable in process mapping and team dynamics) and a team of knowledgeable employees to work as a group to develop the current state process using the sequence laid out in the following list. Start the analysis at the 100,000-foot level, and then repeat the process for the identified 50,000-foot-level diagrams as needed for review of the situation and to clarify the gap identification.

- *100,000-foot level.* Low level of detail; provides a summary of a process at the system level. Captures the major phases of the value chain processes.
- *50,000-foot level.* Medium level of detail; breaks down a high-level process phase into the core process steps.
- *Ground level.* Captures process detail at the subgroup or even the individual level.

Figure 5-2 shows the high-level flow for this analysis effort. The steps in process mapping and improvement are as follows:

1. Identify the beginning and end of the process or the process segment associated with the need for improvement or gaps in performance.
2. Determine the key process participants and/or stakeholders.
3. Document the key process steps.

   3.1. Try to keep diagrams as simple and clear as possible.

   3.2. Make sure to document the process *as it actually is,* not how it should be.

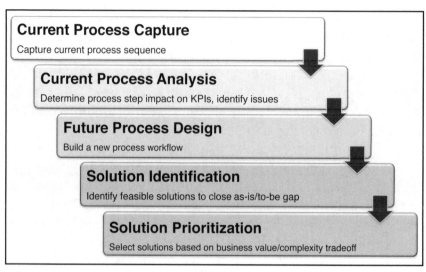

FIGURE 5-2.   Process mapping and improvement workflow.

4. Determine the activities and cycle times and other relevant KPIs. Business processes consist of three primary types of activities:

   o   *Value-adding activities* provide products or services to the customer.

   o   *Handoff activities* move work across organizational boundaries.

   o   *Control activities* provide a standard and measurement checkpoints in a process.

5. Determine the work and cycle time for each step in the process.

   o   Work time = labor hours directly attributable to process step

   o   Cycle time = elapsed time from step beginning to step end

6. Additionally, identify any relevant KPIs of the process. Examples could include:

   o   Number of units processed daily/monthly/yearly by the organization

   o   Number of units processed per person

- o Cost to process the unit or service
- o Customer or stakeholder satisfaction with the process
7. Start at measurable gaps in performance.
   - o KPIs are measurable pains when they are not meeting goals.
   - o Pains have *no quantitative value* if they cannot be measured.
8. Complete the process map with pains and indicators shown directly on it.
9. Vet the current state process map with the team.
10. Walk the process as documented, identifying pains and problems, stakeholders, and so on.
11. Future state process—design goal: to address key areas of improvement identified during the current state analysis.
12. Identify solutions:
    12.1. Brainstorm solutions or process improvements and work with the group to narrow the list of solutions to those most valuable and most feasible.
    12.2. Select and prioritize process improvements for inclusion in the "to-be" process.
    12.3. Narrow the list of ideas, ranking them against the impact to address the problem and the overall implementation effort, considering such things as:
    - o Ballpark cost, ballpark level of effort, degree to which the solution addresses the problem, degree of change management/estimated cultural acceptance
    - o Type of value it would represent
    - o Degree of difficulty to implement this solution
13. Develop the to-be process map using the subject matter experts and solution ideas generated.
14. Document the anticipated benefits of the process changes.
    - o Use a level of detail necessary to fully communicate process pains or process improvement benefits.
    14.1. Get early consensus for the "to-be" process maps.

15. Tailor the complexity of the implementation plan to the business's needs and expectations.

16. Execute the implementation plan.

SIMULATION

BPM provides the tools to create the model or flowchart of your critical business processes. It allows you to understand how the processes operate in your business. Once you have applied this methodology, the next logical step is to use this knowledge and add simulation to your tools.

Today's technology has provided an even more effective and useful tool: simulation modeling. If a picture (flowchart) is worth a thousand words, then one that logically simulates tasks and collects data has to be worth a million. Simulation models have the capability of considering complex interrelated tasks and structurally projected outcomes in a matter of seconds, providing users with validated, and reliable, results.

Simulation provides a less expensive means of experimenting with a detailed model of a real process to determine how the process will respond to changes in its structure, environment, or underlying assumptions. As you attempt to improve or streamline your processes to meet your current business conditions, simulation provides you with another tool. It allows for a better understanding of process changes with the goal of improving performance, while experiencing no disruption of normal business activity. Simulation modeling provides a structure and a method to evaluate, redesign, and measure process changes, while minimizing the amount of time spent, the resource utilization, and the risk.

When you combine BPM with process simulation, you have a powerful means by which to design, evaluate, and visualize new or existing processes without running the risks associated with conducting tests on a functioning business process. Dynamic process simulation allows organizations to study their processes from a systems perspective, thereby affording them a better understanding of cause and effect, in addition to predicting outcomes. The strengths and capabilities of simulation make it an ideal tool for process redesign. It aids in evaluating, redesigning, and measuring, with the following benefits:

- Improved cycle time
- Effective use of resources/improved productivity
  - o Non-value-added work
  - o Wait times/queues
- Identification of critical process parameters
- Modification of critical process parameters
- Improved critical-to-quality parameters
- Improved customer satisfaction

During a process design or redesign project, simulation can assist in the following areas:

- Feasibility analyses
- Examining the viability of new processes while looking at various constraints
- Cost-benefit analysis or process evaluation
- Exploring the possibilities for the system in the future state
- Examining the performance metrics of a system in both current and future states
- Prototyping, once a future state vision for a redesigned process is generated
- Modeling implementation planning, risk assessment, and process design
- Disseminating information about the newly reengineered process to the organization

Simulation can assist in creative problem solving by providing a less expensive way of testing ideas. Fear of the high cost of failure prevents people from coming up with creative ideas.

Simulation allows for experimentation and testing, followed by selling the idea to management. The use of simulation can help to predict outcomes. For example, it could help in predicting the response to increases or decreases in market demands placed on a business process, analyzing how the existing infrastructure can handle the new demands placed on it. Simulation can help determine how resources may be effectively and efficiently allocated.

Conventional analytical methods, using static mathematical models, don't effectively address variation because calculations are made from constant values. Simulation looks at process variation, by taking into consideration the interaction among components, the appropriate statistical distribution, and time. It allows for a more sophisticated analysis of all the variables.

This approach used in simulation promotes total solutions by modeling the entire process. It provides insight into the capabilities of the process and the effect process changes will have on the inputs and outputs. The biggest bonus comes from using a simulation model for experimenting with process parameters without making changes to the functioning process as it exists. It allows practitioners to test more alternatives, lowering risk and increasing the probability of success, and it generates information for decision support.

The biggest plus in applying simulation can be cost effectiveness. As organizations try to respond quickly to rapid changes in their markets, a good working simulation model can be an excellent tool for prototyping and evaluating rapid changes. For example, a sudden change in market demand for a product or service can be modeled using a validated process model to determine whether the existing process can meet the new demand, either up or down.

Another use for simulation can be validating performance metrics. For example, the aim of a key business process may be to satisfy a customer within a specified time frame. Using a simulation model, this requirement could be translated to the time required to respond to a customer's request, which can then be designated as a key performance measure resulting in customer satisfaction. Simulation can help test the tradeoffs associated with process designs and allow for further analysis on parameters such as time to market, service levels, cycle time, production costs, inventory levels, and staffing levels. Simulation can provide a quantitative approach to establishing key performance metrics.

Simulation can also provide an effective communication tool. It can be used to introduce a new or redesigned process in a dynamic fashion. Using simulation to display the functioning process provides a powerful means of explaining the function of various components to those who will work in the new process; in turn, it helps them understand how it works and fits together in the big picture.

Businesses don't lose customers over typical performance; it is the extremes and variations in performance that make customers unhappy—for example, your business commits to answer the phone on or before the third ring 95 percent of the time, and it typically takes six to eight rings for a customer's call to be picked up. Simulation modeling is the most effective way to do the type of analysis required. It allows you to develop a staffing plan that enables you to answer the phone in three rings 95 percent of the time.

PROCESS REENGINEERING[1]

Business Process Reengineering (BPR) is radically redesigning an organization's existing processes. BPR, however, is an approach for redesigning the way work is done to better support the organization's mission and reduce costs. Reengineering typically starts with a high-level assessment of the organization's mission, strategic goals, and customer needs. Once the organization rethinks what it should be doing, it proceeds to rethink how best to radically redesign the processes.

Within the framework of this overall assessment of mission and goals, reengineering focuses on the organization's business processes—the steps and procedures that govern how resources are used to create products and services that meet the needs of

---

[1] Information in this section was compiled from Wikipedia and *GAO Business Process Reengineering Assessment Guide* (U.S. General Accounting Office, Version 3, May 1997).

particular markets, market segments, or customers. As a structured ordering of work steps across time and place, a business process can be decomposed into specific activities, measured, modeled, and improved. It can also be completely redesigned or eliminated altogether. Reengineering identifies, analyzes, and redesigns an organization's core business processes with the aim of achieving dramatic improvements in critical performance measures, such as cost, quality, service, and speed.

Reengineering recognizes that an organization's business processes are usually fragmented into subprocesses and tasks that are carried out by various functional areas within the organization. As discussed earlier, there is no single process owner who is responsible for the overall performance of the entire process. Reengineering maintains that optimizing the performance of subprocesses can result in some benefits but cannot yield dramatic improvements if the process itself is fundamentally flawed. For that reason, reengineering focuses on radically redesigning the process as a whole in order to achieve the greatest possible benefits to the organization and its customers. This drive for realizing dramatic improvements by fundamentally rethinking how the organization's work should be done distinguishes reengineering from process improvement efforts that focus on functional or incremental improvement. Table 5-1 shows some of the areas of comparison between process improvement and process reengineering.

Process redesign using BPR principles generally starts with a clean slate. It is a great deal more complex, time consuming, and, above all, risky. BPR does not mean just change but, rather, dramatic change. What constitutes this drastic change is the remodeling of organizational structures, management systems, employee job responsibilities, performance measurements, incentive systems, skills development, and the application of IT. BPR can potentially impact every aspect of the way business is conducted today. Change on this scale can cause results ranging from enviable success to complete failure. There has been a great deal written in this area, and it is strongly recommended that before you embark down this path you apply some due diligence and investigate the successes and failures. Another effective method used to reengineer business processes is a Design for Six Sigma approach, which is discussed later in this chapter.

TABLE 5-1. Differences between Process Improvement and Process Reengineering

	PROCESS IMPROVEMENT	PROCESS REENGINEERING
LEVEL OF CHANGE	Incremental	Radical
STARTING POINT	Existing process	Clean slate
FREQUENCY OF CHANGE	Continuous	One-time
TIME REQUIRED	Short	Long
PARTICIPATION	Bottom-up	Top-down
TYPICAL SCOPE	Narrow, within functions	Broad, cross-functional
RISK	Moderate	High
PRIMARY ENABLER	Technology	Technology
TYPE OF CHANGE	Cultural/structural	Cultural/structural

SIX SIGMA

The roots of Six Sigma go back nearly 60 years, to the post–World War II Japanese management breakthroughs and extend through the "Total Quality" efforts in the 1970s and 1980s. Six Sigma is a methodology that provides an organization with the tools to improve the capability of its business processes. This increase in performance and decrease in process variation can lead to defect reduction and significant improvement in profits, employee morale, and quality of product. Simply stated, the goal of Six Sigma is to help people and processes deliver defect-free products and services.

Three key characteristics separate Six Sigma from other quality or process improvement initiatives. Six Sigma:

1. Is a customer-focused approach
2. Yields and tracks the returns on investment from projects
3. Changes how management functions

Six Sigma is just as much a business initiative as it is a quality initiative. It's about preparing the entire organization to meet the changing needs of markets, technologies, and customers.

Six Sigma is a well-structured, data-driven methodology for eliminating defects, waste, or quality problems in all kinds of businesses—manufacturing or service, large or small—by making incremental process improvements. The methodology is based on the combination of well-established statistical process control techniques, data analysis methods, and the systematic training of personnel at every level in the organization involved in the processes targeted.

The following outlines the six major themes of Six Sigma.

1. *Focus on the customer.* Customer focus becomes the top priority. Improvement projects are measured by their impact on customer value and satisfaction.

2. *Fact- and data-driven management.* Rather than act on opinions and assumptions, Six Sigma clarifies what measures are necessary to gauge organizational performance, then gathers data and analyzes key variables. By doing this, problems can be more effectively defined, analyzed, and resolved permanently. These essential questions need to be answered to support data-driven decisions and solutions:

   o What information/data do I really need?

   o How do I use the information/data to its full potential?

   o Is my measurement system accurate enough for what I am doing?

3. *Processes are the key.* Improving processes is the way to build competitive advantage in delivering value to customers.

4. *Proactive management.* This methodology focuses on problem prevention and asking why things are done a certain way, rather than resorting to a reactive mode.

5. *Collaboration without boundaries.* By breaking down the barriers across organizational lines and between departments, Six Sigma works to negate competitiveness and miscommunication so that everyone can be working toward providing value to the customer.

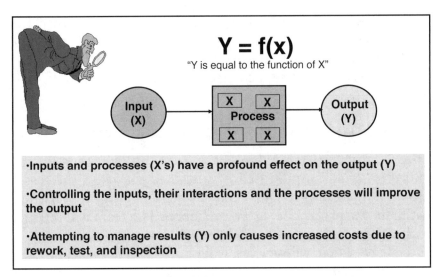

FIGURE 5-3. The essence of Six Sigma.

6. *Drive for perfection and tolerate failure.* If people see possible ways to come closer to perfection but are too afraid of the risks and consequences involved, they will never try. Any organization with Six Sigma as a goal must continue to strive for perfection, while being willing to manage and accept occasional setbacks.

There is a basic mathematical concept that states that Y (results) is equal to the function of X (inputs and processes). In business, we often expend all of our attention on the output (repair, redo, recycle), which always adds time and money.

Six Sigma forces us to look at the inputs and processes (function of X) as they affect the output (Y). Figure 5-3 shows a graphical representation of this concept. If we are able to control inputs and improve processes, the outcome will automatically improve without redoing anything.

When people say they are using Six Sigma, what do they mean? Which methodology are they using? The majority of the time they are using the DMAIC (Define, Measure, Analyze, Improve, Control) methodology because they already have existing products or services that are not performing up to expectations. Because it is more widespread, this chapter discusses DMAIC first, followed by a brief discussion of the other methodology—Design for Six Sigma (DFSS)—which can be used to design or redesign complete processes.

THE DMAIC PROBLEM-SOLVING PROCESS

DMAIC teams are created to solve process problems and to capitalize on opportunities. Led by a trained specialist, the teams usually contain 3 to 10 members representing the groups participating in the process being worked upon. The team will interact with the larger organization, interview customers, gather data, and talk to people whose work will be affected by the team's recommendations.

Steps in the DMAIC Life Cycle

- *Pinpoint and select the project.* Management reviews a list of possible projects and selects the most promising to be tackled by the team.
- *Structure the team.* A specialist (Black Belt or Green Belt) is selected. Management selects team members who have good working knowledge of the process but are not resistant to change.
- *Create the charter.* The charter includes the reason for pursuing the project, the goal, a basic project plan, the project scope, a business case, and a review of roles and responsibilities. It is the key document that provides the written guide to the problem and is drafted by the Champion and the team.
- *Train the team.* Training is carried out for the DMAIC tools and processes involved.
- *Perform DMAIC and implement solutions.* The team must develop pilots, training material, project plans, and procedures for their proposed solutions. The team is also responsible for implementation, measurement, and

monitoring of the results for a meaningful set period of time.

- *Hand off the project.* Once the work of the DMAIC team is finished, the solution is handed off to the process owner. Often, some of the team members continue to work in the affected area to help manage the process or solution and drive it toward continued success. Other team members return to their regular jobs, but not without having learned some new skills and gained experience that can be applied elsewhere.

The DMAIC Problem-Solving Model

1. *Define the problem.* The team creates the charter that defines its focus and is the blueprint for the project. Next, the team identifies the customer. The team must listen to the customer and translate the customer's language into meaningful requirements. Then, a high-level diagram of the process is created. This diagram usually shows 5 to 10 major steps describing the current process. This diagram is meant to get all members of the team on the same page and to set the stage for the next major step in the process.

2. *Measure.* This step has two main objectives: to gather data to confirm and quantify the problem or opportunity and to begin searching for facts and numbers that offer clues about the causes of the problem. Also, the team looks at and analyzes the data collection system to ensure that the accuracy is statistically good enough to make the critical measurements. Three measures can be taken to determine the source of the problem:

   o *Input.* Anything that comes into the process to be changed into an output. Bad inputs lead to bad outputs.

   o *Process.* Anything that can be tracked and measured. These things usually help the team pinpoint and track the problem.

o   *Output or outcome.* The end results of the process. End results focus on immediate results, such as deliveries, defects, or complaints, and more long-term impacts such as profits and satisfaction.

The team's first priority is the output measure. This baseline measure is the data used to complete the charter. Then a few input measures are targeted to begin getting data on potential causes. Once the team members have determined what to measure, they create a data collection plan, whereby they collect the data that is required. An initial sigma measure for the process being fixed should then be established.

3. *Analyze.* The team determines the root cause of existing problems. One of the principles of good problem solving is to consider many types of causes and to prevent biases or past experience from clouding judgment. Some of the common cause categories to be explored are procedures, techniques, machines, materials, measures, environmental elements, and people. The analysis begins by combining experience, data/measures, and a review of the process, and then forming an initial theory of the cause. The team then looks for more data and other evidence to see whether it fits with the suspected cause. This cycle of analysis continues, with the theory being rejected or refined until the true root cause is identified and verified with data.

4. *Improve.* A good team leader must recognize that many teams are tempted to jump right to this step from the start of the project. Solutions have to be carefully managed and tested. Changes have to be sold to the organization members whose participation is critical. Data must be gathered to track and verify the impact of the solution.

5. *Control.* The main goal of this step is to prevent the people and processes from reverting to old habits. There has to be long-term impact on the way people work. Specific control tasks that DMAIC teams must complete are:

o   Creating a monitoring process to keep track of the changes they have set out o   Developing a response plan for dealing with problems that may arise

o   Selling the project through presentations and demonstrations

o   Focusing management's attention on a few critical measures that provide current information on the outcomes of the project

o   Ensuring support from management for the long-term goals of the project

o   Handing off project responsibilities to those who do the day-to-day work

The Six Sigma methodology for process improvement is DMAIC (Define, Measure, Analyze, Improve, and Control). This methodology is consistently used by organizations in all industries. Figure 5-4 provides a summary of some of the tools that may be used during this process improvement methodology.

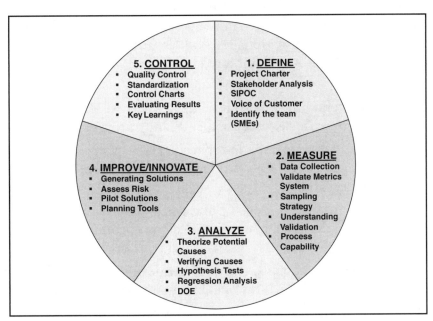

FIGURE 5-4.   DMAIC cycle tool summary.

DESIGN FOR SIX SIGMA (DFSS)

DFSS is the methodology that is most commonly used by organizations that are redesigning or creating a new process. The steps or phases in DFSS are not as clearly defined as they are in DMAIC. DSFF is often defined differently by each organization that uses it. Producing such a low defect level for product or service launch means that customer needs and expectations must be completely understood before a design can be completed and implemented. There are many different acronyms that are used in performing DFSS. This section focuses on the most common ones.

Process design or redesign using Six Sigma is generally referred to as Design for Six Sigma (DFSS) or DMADV (Define, Measure, Analyze, Design, Verify). The objective is not to fix the process, as you would with DMAIC or Lean principles, but to replace it with a new and greatly improved one. The concept here is very similar to the concepts of process reengineering; that is, the process is beyond repair, and it needs replacement or at least a major redesign.

Process design/redesign is the creation of a new process to achieve exponential improvement and/or meet the changing demands of customers, technology, and competition. It must handle totally dysfunctional processes and reengineer them. DMADV is the most common road map followed for DFSS:

- *Define*. Define the goals of the design activity that are consistent with customer demands and enterprise strategy.
- *Measure*. Measure and identify critical-to-quality characteristics (CTQs), product capabilities, production process capability, and risk assessments.
- *Analyze*. Analyze to develop and design alternatives, create high-level design, and evaluate design capability to select the best design.
- *Design*. Design details, optimize the design, and plan for design verification. This phase may require simulations.
- *Verify*. Verify the design, set up pilot runs, implement the production process, and hand it over to the process owners.

LEAN PRINCIPLES

The core concept of Lean principles is to maximize customer value while minimizing waste. *Lean* means creating more value for customers with fewer resources. A lean organization understands customer value and focuses its key processes to continually increase it. The ultimate goal is to provide as close to perfect value as possible to the customer through the value creation process that has zero waste.

Process improvement using Lean concepts is a set of methods that focus on speed, efficiency, and elimination of waste. The goal is to maximize process speed by reducing waste. Waste is defined from the understanding of value from the standpoint of the customer. The key underlying concepts of the Lean approach are defined by the following five principles.

Five Lean Principles

1. Specify value from the standpoint of the customer—what is the customer willing to pay for?

2. Identify all steps in the value stream and eliminate those that do not create value. Identify the value-added, non-value-added, and business non-value-added work in the process.

3. Make value-creating steps occur in tight sequence.

4. As flow is introduced, let customers pull value from the next upstream activity.

5. As value is specified, value streams are identified, wasted steps are removed, and flow and pull are introduced, begin the process again and continue it until a state of perfection is reached in which perfect value is created with no waste.

In applying lean principles, you need to identify process steps and eliminate those that do not create value, ensuring that value-creating steps occur in tight sequence. Let customers pull value by producing at the rate of demand. Then do it all again.

Lean tools include several different things. The most important tool and the most closely related to BPM is *value stream map-*

ping. This is a paper-and-pencil tool (which can later be stored electronically) that helps you to see and understand the *flow* of material and information as a product or service makes its way through the value stream. Looking at a process on a value stream map (VSM) allows you to see the process from end to end and identify steps where no real value is added (bottlenecks, delays, excess inventory, defects, and so on).

The goal of value stream mapping is to identify waste and provide a road map and an ongoing systematic approach for improvement in the value stream—it enables and optimizes flow. The originating VSM of a given process or product, often referred to as a *current state value stream,* becomes the baseline for improvement initiatives that eliminate non-value-added and wasteful activities and enable flow.

A VSM includes the materials, information, and processes required to design, produce, and provide a product or service for a customer. A VSM includes all elements (value-added, business non-value-added, and non-value-added) that occur to a product or service from its inception through delivery to the customer.

Value-added steps in a process are exactly that: work that is expended on a product or service for which a customer is willing to pay. Non-value-added work is the waste in the process. Business non-value-added work is the steps and activities in a process that the business is required to perform because of regulatory or business record-keeping requirements. In other words, this is work expended that does not add value to a product but must be done to meet other business requirements.

The VSM helps you visualize and optimize the whole rather than individual parts. It links material and information flows and shows total lead time, it identifies the ratio of value-added to non-value-added time, it provides a blueprint for implementation, it provides a framework for more quantitative tools, and it ties together Lean concepts and techniques.

Lean is all about learning how to reduce waste, or non-value–added, steps in a process. Typically 80 to 95 percent of lead time is non-value-added, or waste. Eliminate the waste in your process, and you will produce services more efficiently and more profitably. The acronym WORMPIIT is used to help you remem-

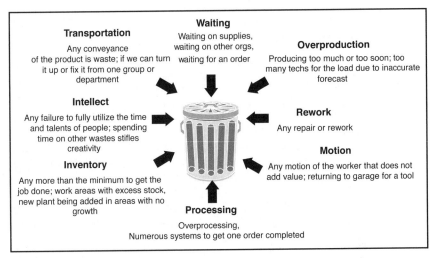

FIGURE 5-5.   The eight types of waste—W.O.R.M.P.I.I.T.

ber the eight types of waste: waiting (W), overproduction (O), rework/defects (R), motion (M), [excess] processing (P), inventory (I), intellect (I), and transportation (T). See Figure 5-5 for examples of each category of waste.

KAIZEN

Another methodology, occasionally connected to Lean principles, is kaizen. In Japanese, the word *kaizen* means "change for the better." A kaizen event is intended as a relatively quick method to improve a process. The technique was developed for manufacturing, but it is now applied to service industries as well. It also takes advantage of the knowledge of the participants in the process to perform the analysis and recommend the changes and improvements. People who do the work are the ones who truly know how to improve it.

The cycle of kaizen activity can be described as follows:

1. Standardize an operation in the process.

2. Measure the standardized operation (find cycle time and amount of in-process inventory).

3. Gauge measurements against requirements—look for the gaps in performance.

4. Innovate to meet requirements and increase productivity.

5. Standardize the new, improved operations.

6. Continue the improvement cycle indefinitely.

Table 5-2 provides a quick overview of the differences between the Lean approach and Six Sigma (DMAIC). The Lean approach does not delve as deeply into a problem and can be accomplished in a shorter period of time. Six Sigma tends to dig very deeply into a specific problem and can be time consuming due to the level of research and analysis required.

TABLE 5-2.   High-Level Comparison of Six Sigma (DMAIC) and Lean Principles

SIX SIGMA	LEAN
• Focus on subprocesses	• Remove waste, rework
• Remove variation	• Improve flow, velocity
• Research process problems	• Fast results
• Design more capable processes	• Focus on system
"An inch wide and a mile deep"	"A mile wide and an inch deep"

LEAN SIX SIGMA

Lean Six Sigma is a process improvement methodology that combines tools from both Lean and Six Sigma methodologies. The Lean methodology focuses on speed, and traditional Six Sigma focuses on quality. The combination of the two methodologies is intended to meld the two different tool sets into one, resulting in better quality faster. Figure 5-6 gives the highlights of the combination of the Lean and Six Sigma methodologies. Lean principles and DMAIC are directed toward redesigning and improving processes to meet the expectations of customers and to remain competitive. The Lean approach focuses on improving

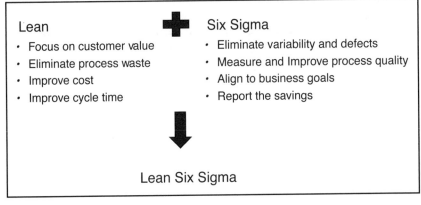

FIGURE 5-6   Lean Six Sigma.

the flow of products or services to the customer. DMAIC focuses on the quality of the product or service by reducing variation in the features and functions that the customer values. Nonvarying, predictable, reliable inputs to the process are necessary for process stability—a requirement before Lean principles can be rigorously applied to improve flow.

CONCLUSION

These process improvement tools are examples of how you can make the changes required in your business processes when they are not meeting the needs of the organization. They need to be applied by knowledgeable practitioners. As you look at the differing approaches in each of these methodologies it should be evident that each suits a specific purpose and it is not a one-size-fits-all approach to process improvement. These methodologies equip us with a toolbox from which to select the right tool for the right job.

PROCESS MANAGEMENT

PROCESS MANAGEMENT FUNDAMENTALS

Process management relates to the activities of planning and monitoring the performance of the organization's business processes. Process management is the application of knowledge, skills, tools, techniques, and systems to define, assess, improve, and manage the value stream processes with the goal of meeting customer requirements profitably. Process management is different from program management because the latter is concerned with managing a group of interdependent projects. However, process management includes program management. In project management, process management is the use of a repeatable process to improve the outcome of the project. Table 6-1 summarizes the similarities and differences between projects and processes.

TABLE 6-1. Comparison of Processes and Projects

PROCESSES	PROJECTS
Continuous	One-time set of events
Triggered by events	Defined start and end dates
Set of activities that evolve over time	Specific set of tasks
Support business transactions	Specific set of objectives
Replicated and repeated	Each one unique
Resources allocated to perform the activities	Resources allocated to project tasks
	Resources reallocated when it ends
Continuous improvement	Changes are avoided

Process improvement is typically executed by the use of projects that employ one or more of the methodologies outlined in Chapter 5. The evaluation of a project is done at key milestones, and when the project ends, an evaluation is performed to determine whether the objectives, deliverables, or goals have been met. The evaluation of a process is done frequently, at specified intervals during the process, and over different time frames, against a multiple and changing set of objectives.

MANAGING THE INTRODUCTION AND PERFORMANCE OF A NEW PROGRAM—BPM

As the management team better understands the key essentials in Business Process Management and makes the business decision to proceed, the organization needs to roll out this new program. BPM program rollout best practices include:

- Clearly defined roles and responsibilities
- A process measurement system
- Monitoring of process performance
- Cross-functional process management
- Change management for processes
- Defining, prioritizing, and managing of improvement initiatives
    - o Governance—decision making
    - o Process portfolio management
    - o Oversight of improvement initiatives

The following approach to applying BPM enables an organization to achieve the desired results. The most salient aspects include:

- A business process management program that is simple and credible to an organization's leadership team—from the senior management through all levels in the entire organization

- Incorporation of process rigor and an understanding of what is required to transform an organization and its people
- Enabling of a focus on business strategy and measurement to work as a catalyst for strategy execution

The primary objective of phase 1 of this program rollout is to introduce the fundamentals of BPM to the organization and create a high-level value stream map for the business. The next step is to identify key process indicators (KPIs) for those processes. Using the data collected, gaps in performance can be identified. Once this has been successfully completed, the core processes can be documented in more detail. Phase 2 identifies a core process that needs improvement, using the process maps, KPIs, and gap analysis.

The program assumptions are that senior management is fully committed to this endeavor and the business is developing a dedicated team that will be engaged in this effort from day one so team members can learn and develop the capabilities of the methodologies for evaluation of current business processes. They also will bring an invaluable understanding of the functional areas and current processes. The team will develop and document the value chain and the "as-is" state of the current business processes.

The functional organization management is responsible for making employees at all levels of the organization available for participation, informing them of the project, for providing information about the organization and past performance metrics to evaluate competencies; for reasonable access to senior management for ongoing progress reports, discussions, and problems; and for coordinating workflow and priorities to allow the project to be completed within its allotted time frames.

APPROACH

The approach to instituting the BPM methodology in an organization involves using real-world experience and practical methods to create a BPM culture throughout the organization.

PHASE 1

Stage 1: Initiate Project—Identify the Value Chain Processes and Build a Process Inventory

In preparation, it is helpful for the facilitator to have the following information (if available) on or before meeting with the senior management team.

- A statement of key strategic initiatives
- A list of key organization goals, objectives, and/or targets
- A copy of the organization chart (annotated with headcount numbers, if available)
- A summary of recent customer satisfaction data
- An employee satisfaction survey summary
- A summary of recent process improvement work (and results)
- Any other documents that would help the team understand the general business situation

Conduct Interviews
- Interview the members of the senior management team.

Develop Draft Materials
- Prepare a process inventory (value chain and support processes).
- Create a relationship map for the value chain.
- Identify the key process metrics (KPIs).
- Generate estimates of current performance.
- Jointly identify potential benchmarking/best practices partners.

Conduct Session One with the Business Leadership Team
- Clarify assumptions.
- Agree on strategic objectives.
- Refine the process inventory.

- Refine the process relationships.
- Validate and refine the key process metrics.
- Discuss the size of any performance gaps.
- Identify draft priorities.

Complete Interim Work
- Refine the key documents based on the first session with the leadership team.
- Collect additional data on current performance if needed.

Conduct Session Two with the Business Leadership Team
- Review the outputs from the first session.
- Agree on the value chain (core process) definitions.
- Agree on the support process definitions.
- Identify the top improvement and management priorities.
- Agree on ownership of the value chain/core processes and support processes; identify the process owners.
- Develop the process improvement and management plan.
- Develop the communication plan for BPM rollout.

Stage 2: Create As-Is Process Maps

The approach in Stage 2 is that only the value chain/core processes are mapped at this time. Expect between five and ten value chain processes; this is typical for most organizations.

Complete Prework
In preparation, it is helpful to complete the following information-gathering steps:

- Identify the subject matter experts (SMEs) for each value chain process, with the assistance of the process owners.
- Agree upon the boundaries (start and end) of each process.
- Identify the suppliers to the process to be analyzed.

- Identify the customers of the process to be analyzed.
- Identify any support partners for the process.

Create the "As-Is" Value Chain Process Map

- Convene the previously identified team for a one-day session.
- Collaboratively create the medium-level "as-is" value chain process map (15 to 25 steps).
- Identify disconnects (problems, bottlenecks, costs, cycle time issues, etc.) and KPIs for the current process.
- Reach team consensus on the process map, disconnects, and measurements needed.
- Group the disconnects by common threads in order to identify potential improvement projects.

Document the "As-Is" Value Chain Process Map

- Convert the team's work to an electronic document (using Visio, Igrafx, etc.). Convert disconnects to a Word file and identify improvements.
- Transfer ownership of the electronic files to the BPM team and process owner.

Validate the Results of the "As-Is" Documented Process with the SMEs and Process Owner

- Conduct a follow-up meeting to review results with the team and process owner.
- Walk through the process if possible.
- Finalize the "as-is" process documentation and identified KPIs.

Repeat the procedure for each identified value chain process (typically, 5 to 10 value chain processes).

Conduct Session Three with the Business Leadership Team

- Review the outputs of the "as-is" process maps.
- Validate the value chain/core process definitions.

- Validate the top improvement and management priorities.
- Validate ownership of the value chain/core processes.
- Complete the process improvement and management plan.
- Develop the communication plan for the next phase.

PHASE 2

The initial rollout of any new program into an organization is critical. If it is not done well and successfully, the program will be doomed to failure. Phase 2 is where the organization initiates the process improvement aspect. Because this aspect is the most visible to the employees, success here is critical for the program to be taken seriously. Following are some things to consider in the selection of the first project:

- Think big—envision the rollout of the program.
- Start with a well-chosen initial project to build expertise and acceptance.
- Start small—choose a project with these characteristics:
  o Addresses an obvious pain point/problem to gain credibility with employees.
  o Has an achievable scope—don't try to fix the biggest problem first.
  o Seems feasible, with a high likelihood for success.
- Demonstrates a return on investment (ROI)—proved by cost savings. Work fast—do it quickly.

There are some other factors to consider in order to be successful with your organization's BPM rollout. It is a good strategy to take an incremental approach to the business and information technology benefits by identifying the right pilot project, setting realistic expectations, and establishing cooperation between the business/process owner and the IT department. There needs to be a structured approach, like the one outlined earlier, with a focus on strategy (both the business strategy and the BPM rollout strategy). The team needs to be formed early, with a focus on initial planning, process, and architecture. The organization needs to establish strong governance.

Some key roles relating to governance need to be defined and filled. First, create a steering team that is cross-functional and has decision-making authority. Identify a project sponsor who has both an interest in the process and the authority to make the required changes. A process owner needs to be identified, and it should be the person who has the most to gain or lose from the execution of this process. Also required is a project facilitator—someone who has been trained in the BPM methodology and is knowledgeable in process improvement methods. If your organization lacks these resources, then you can turn to outside consultants until an internal resource has been developed through training and experience.

CHAPTER 7

A PROCESS-MANAGED ENTERPRISE

When an organization decides to deploy a strategic initiative, whether the approach is called Business Process Management (BPM) or something else, it must stay focused on the primary objective: making improvements in organizational performance and sustaining those gains. To ensure the initiative's success, the organization's executive team must design the initiative to take into account its people, history, and culture. The key characteristic of the approach is that it must be systematic. Everyone in the organization must know and understand what is going to happen.

The basic thing to consider when implementing this new strategy is this: Do not try to be another company. Study companies that have had success with process management, and look for examples of "best in class." The executives should pick those strategies or elements of strategy that will work in their organization and internalize them. They should feel free to merge strategies into something that is unique and will be effective in their organization. A company must be careful not to exclude things that it has utilized effectively.

The message is twofold. First, companies must adopt an organization-wide strategy to drive the changes in all areas of the organization. This should be the most important aim of management. Second, companies should build their strategies utilizing the best of the available tools and methods, as well as new methods, based on the knowledge of their own management teams.

THE KEY ELEMENTS OF A PROCESS-MANAGED ENTERPRISE

A process-managed organization is structured, organized, and managed in terms of its core business processes. It is measured and rewarded based upon process performance. The key elements of a process-managed business are:

- A managed portfolio of processes including process owners
- Processes aligned with company strategy
- A change management process
- A method to develop new processes
- A systematic method of analyzing the impact of new and existing processes
- The ability to customize and combine processes
- Methods to analyze and constant improvement processes
- Process models for simulation

In order to fully achieve the value of BPM, an organization needs to change the culture just as it may have done with respect to other initiatives. The attributes of this organization include a continuous process improvement strategy, process-driven budgeting and resource allocation, process execution that directly supports strategy execution, and process architecture that ties to the information technology architecture.

A process-managed enterprise relies on BPM as an approach to manage performance, focusing on the continuous improvement of the value chain, or core, processes. The key elements of BPM are the clearly defined and documented processes, explicit process performance objectives, process owners, and ongoing measurement, resulting in continuous improvement of the core processes.

Process ownership is an absolute essential. The process owner is the central figure in any process-managed enterprise. The process owner must be at a senior management level in order to have the authority and responsibility for the performance and improvement of these critical cross-functional core processes. The process owner is responsible for:

- Process definition
- Process implementation
- Process monitoring
- Process control
- Process analysis
- Process design
- Process deployment
- Performance management
- Financial analysis
- Change management

CREATING A ROAD MAP TO A PROCESS-MANAGED ENTERPRISE

Start by identifying the key stakeholders. They are the senior management team. Next, it is essential to define BPM in terms and within a context that these managers will understand. The financial benefits need to be clear and understandable. Determine how much value BPM will deliver over time. Provide gap analysis and scenarios for improvement, including the benefits. Develop a long-term (two- to three-year) road map for implementation.

Creating a *center of excellence* (COE) for BPM is one way to initiate the change to a process-managed enterprise. A center of excellence for BPM is also sometimes called a *business process management organization*. This organization creates the strategic focus for becoming a process-driven and process-managed enterprise, and it is where skilled BPM professionals perform their responsibilities, including the following:

- Ensuring the definition of all core processes in an organization's value chain and how these processes fit together to create the value chain
- Assigning core processes to executive process owners and facilitating a formal dialogue between the owners and BPM professionals
- Establishing a disciplined and measurable process management program
- Communicating throughout the organization a shared understanding of the key core business processes that deliver value to customers
- Facilitating a cultural change to a process-managed enterprise

The implementation of BPM in an organization introduces many things to the business. The most critical aspect is *process governance*. The key aspects of process governance are:

- Establishing mechanisms and policies that define who is empowered to make certain decisions about the core business processes
- Establishing mechanisms and policies to measure and control the way decisions are implemented in the core business processes

Process governance is an extension of corporate governance. It defines the decision-making rights associated with the definition and deployment of process improvement initiatives. It includes the mechanisms and policies used to measure and control the way enterprise processes are defined, deployed, maintained, and monitored. Process governance strives to blend the flexibility of process management with the control of traditional IT architectures.

A process governance framework enables an organization to answer the following questions:

- What happens when a process is changed?
- How can you be sure the core process you are executing is of high quality?
- How can you be sure a new core process is compliant with IT, business, and regulatory policies? How can you ensure reliability of a core process?

The plan for executing a process governance strategy has several key elements. They must be agreed on by senior management and then put in place. First, organizational goals and strategies must be used to drive the definition of principles to guide the use of BPM in the organization. Second, the areas of activity and responsibility subject to process governance must be identified. Third, senior management must have clear and agreed-to goals for process governance. Fourth, the organization structure and style must be clearly identified and defined, including who has the responsibility and accountability to make the appropriate decisions consistent with the agreed principles for each area.

Process governance operations or the execution of the process governance strategy can be seen as addressing two primary sides:

- *The process management side.* Where and why should we invest in our processes? Where should the organization allocate resources to produce the greatest return, and how do we measure and ensure that these returns are actually achieved? This side of process governance is primarily a process management responsibility.
- *The center of excellence (COE) side.* What are the rules, guidelines, methodologies, tools, standards, and so on, that should be used in changing business processes? This aspect is primarily a process office or COE responsibility.

Figure 7-1 illustrates the concept of process governance.

The implementation of process governance should be centered on the four critical areas for the organization: people, processes, technology, and services. The mechanism by which to

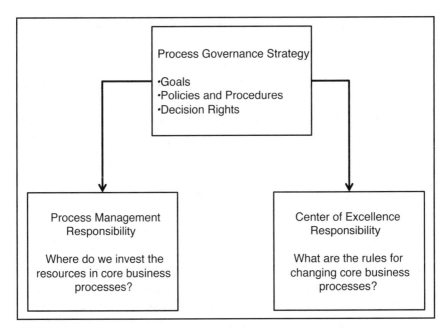

FIGURE 7-1. Process governance.

implement process governance is to establish a center of excellence, generally in the information technology organization. This enables a shared resource and capability center to function as a resource pool as new business application needs arise. The governance implementation needs to be supported by the normal hierarchical organizational reporting structure, and the resources typically overlap with those found in the IT organization.

There are many roadblocks to the adoption of the process management enterprise, among them the lack of understanding of BPM and a lack of understanding of an organization's processes. An organization might not have the methodology to manage and control processes and the supporting architecture. Furthermore, the changeover to process management is impeded by the slow definition of the end-to-end business processes, compounded by attempting to agree with typical industry organization and process terminology, as well as the public versus private nature of business processes.

DESCRIPTION OF PROCESS MATURITY

Process maturity is an indication of how close a core business process is to being complete and capable of continuous improvement through qualitative measures and feedback. For a process to be mature, it has to be complete in its usefulness, automated, reliable in information, and continuously improving. There are numerous qualitative maturity models available in the literature today.

Here is an example of a framework for measuring process maturity. The maturity of a process or activity can be deemed to be at one of six levels, from Level 0 (the least mature) to Level 5 (the most mature). The processes at higher levels also include the features of the lower levels—in other words, the characteristics are cumulative. The ground level is Level 0, where no process exists for the activity. Figure 7-2 shows a graphical representation of a typical maturity model.

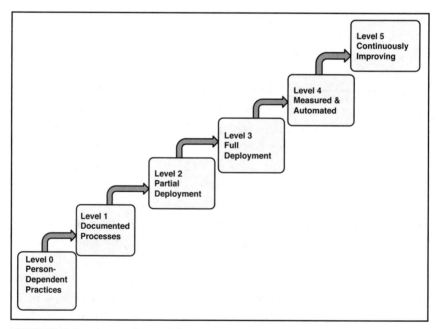

FIGURE 7-2. A maturity model.

Level 0—person-dependent practices. This level is for cases where the activity being performed is not documented. In other words, it is not recorded either in outline or in detail. The activity is entirely person dependent, and the sequence, timing, and result may vary during repetition. This level requires excessive supervision. There is no guarantee of either achieving the desired result or adhering to timelines. The activity is entirely ad hoc, with little communication between functions. The effectiveness of the activity is entirely dependent on individuals. Knowledge transfer may or may not happen if there is any change in the owner of the activity. This is a very dangerous and costly way to run a business.

Level 1—documented process. At this maturity level, there is a document that has been reviewed and approved by the supervisor or the approving authority as the standard process, but it may be doubtful that the activity being performed is as per the document. This may be because of a process drift or some drastic change since the document was drafted. Basically, documentation may exist but it is not maintained or audited for compliance.

Level 2—partial deployment. At this maturity level, the activity that is documented is being deployed, but there is inconsistency in the deployment. The process may not be deployed in totality. That is, it may not be deployed at all the intended locations, or though all functions, or by all the intended participants, or all the activities defined in the process are not being performed. This would mean that the document has not been designed with the required flexibility. There is inconsistency in the results of different local process participants.

Level 3—full deployment. At this level, there is no inconsistency between the documented process and the deployed process. The process documented and deployed caters to all the intended locations, the participants, and all the activities that need to be performed. The process also shows seamless linkage between functional organizations and other business processes.

Level 4—measured and automated. The process has set itself goals such as adherence to key process indicators—quality, cycle times, customer satisfaction, cost, and so on. The process also is being measured against its goals. The process is system driven by enablers such as using technology like enterprise resource planning, customer resource management, or any other custom-made software.

Level 5—continuously improving. The goals set for the process are being analyzed for achievements and improved regularly at this level of maturity. The timelines, cost targets, and satisfaction levels are being achieved regularly, and the targets also are being tightened by using continuous quality improvement techniques such as Six Sigma, kaizen, or Lean principles. The enabling system also is being improved and made error-free by strategies such as *poka-yoke* (mistake-proofing).

From the inception of a BPM initiative through its growth and maturity, the core business processes of an organization usually improve. But this may not be the case for all of the organization's value chain processes. Some may still be at the lower maturity levels. This shows that the organization has not experienced the overall maturity of its processes and application of BPM principles. There also are instances where the maturity level may drop if not monitored or if the documents are not revised according to changes the business undergoes. Thus, a fully established process-managed enterprise may have core business processes that are at different levels of maturity.

THE SEVEN CHARACTERISTICS OF A TRULY PROCESS-MANAGED ENTERPRISE

If you want to become a process-managed enterprise, not only do you need to manage processes as described here, you also need to look at how your business is organized and operates. Research has shown that the common characteristics of process-managed organizations are as follows:

1. *Strategic alignment.* How well has BPM been aligned with organizational goals and objectives?
2. *Governance.* Has accountability for core processes been assigned and a support organization put in place for process management?
3. *Process documentation.* Have the core processes of the organization been defined and documented?
4. *Change management.* Have rules been put in place concerning changes to core processes?
5. *Process performance.* Are performance of the core processes and the maturity of process management being measured?
6. *Process improvement.* Is the organization continuously improving and optimizing the core processes?
7. *Tools and technology.* Are software tools being used to assist in the management of the core processes?

PROCESS-ORIENTED MANAGEMENT

Let's look at evaluating how process-oriented an organization is in terms of its management structure. The following criteria can be used to determine whether your organization's management structure is function based or process based.

- *Function-based organizations* are hierarchical in nature and manage people within typical boundaries relating to task. There is no consideration given to process activities. These are vertical organizations, as opposed to horizontal, or process-based, organizations. These vertical organizations are based on functions, products, operations, and so on. Generally, there is little or no process-based thinking in a function-based organization.

- *Process-thinking organizations* adopt the concepts of process management and attempt to understand how the processes of the organization work together. Process thinking exists and the fundamentals are in place, but the organization is not yet organized around processes.

- *Process-focused organizations* manage the end-to-end core business processes that are tied to functional activities and goals. A process management infrastructure exists; there is a shift to horizontal management. The organization is integrating process thinking with organizational structure.

- *Process-based organizations* manage completely around end-to-end processes. Functional activities are embedded in the processes, and senior executives are the process owners. Horizontal management is in place.

Once you have determined where your organization is currently and where you want to be in the future, you can build an action plan to make this change. Being fully process-based doesn't make sense for every organization. Most organizations that are successfully managing their processes have a mixed model. The objective here depends on the organization, the industry it is in, its culture, and similar factors.

Before taking on a process management initiative, the organization should undertake an assessment of its processes, process management skills, capabilities, and organizational culture, and then determine how it plans to proceed and what pace is practical. This initiative should not be driven by outside forces—for example, that the competition is embarking on process management.

CHAPTER 8

BPM TOOLS AND TECHNOLOGIES

BASIC TYPES OF TOOLS AND TECHNOLOGIES

BPM tools cover a wide spectrum starting with basic mapping tools and including business process analysis tools, enterprise modeling tools, and BPM software suites. This chapter explores what each type of tool has to offer.

- Basic *process mapping tools* are just that—simple software packages, which are inexpensive and enable the user to draw a process map. No analysis capability exists in these packages. Microsoft Visio is one example of this type of tool. Many of the more sophisticated products provide a bridge to import Visio diagrams.

- A *business process analysis* (BPA) tool is stand-alone modeling software that provides the capability to model a process, simulate its performance, optimize it, and design a new process. iGrafx is an example of a BPA tool.

- An *enterprise modeling platform* is an enterprise modeling solution that enables a team to visualize, analyze, document, and optimize an organization's business processes and systems. It enables users to model organizations, processes, data, and systems.

- A *Business Process Management suite* (BPMS) is a software infrastructure product that provides the capability to model, design, deploy, execute, analyze, and optimize end-to-end business processes. It coordinates the flow of tasks to both people and systems and provides for access to resources and the exchange of information among systems, employees, customers, and partners while capturing information about the execution of the process to enable continuous process improvement. BPM suites are provided by many software and technology companies today.

BPM SUITES

Do not confuse the application of a BPMS as "the whole of BPM." A BPMS is a technology enabler—it allows you to put into practice all of the aspects of BPM that have been discussed up to this point. A BPMS encompasses the concept of supporting the application of the BPM approach through enabling technology. A BPMS should enable all stakeholders to have a firm understanding of the core processes and their performance. It should facilitate business process change throughout the life cycle. This assists in the automation of activities, collaboration, integration with other systems, integration of partners through the value chain, and so on. For instance, because of the size and complexity of daily tasks and activities, the use of technology is often necessary in order to model them efficiently. These models facilitate automation and solutions to business problems. They can also become executable to assist in monitoring and controlling business processes. As such, some people view BPM as the bridge between information technology (IT) and the business itself. In fact, an argument can be made that this approach helps bridge the gaps between the functional and technological organizations.

There are four critical components of a BPMS:

- *Process engine.* This component is a robust platform for modeling and executing process-based applications, including business rules.
- *Business analytics.* This component enables managers to identify business issues, trends, and opportunities with reports and dashboards, and react accordingly.

- *Content management.* This component provides a system for storing and securing electronic documents, images, and other files.

- *Collaboration tools.* This component removes intra- and interdepartmental communication barriers through discussion forums, dynamic workspaces, and message boards.

Figure 8-1 illustrates all of the interfaces involved in a BPMS.

The underlying concept of the BPMS is that process logic is externalized. What that means is that it is different from the enterprise resource planning (ERP) and customer relationship management (CRM) systems, which embed the process logic internally in the application. The BPMS takes the activities provided by those applications and composes them in a sequence defined by the process itself, and this allows you to easily change that process logic. The BPMS relies on a new generation of integration middleware that lets users leverage existing systems but compose their functions in user-defined ways.

In a BPMS, the process design looks like a process flowchart that you would have developed as part of your BPM activities.

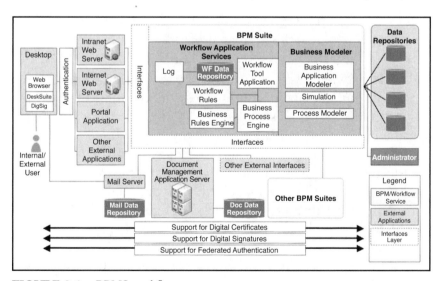

FIGURE 8-1.   BPMS workflow.

This process flow connects the various human tasks and system functions. The benefit comes from the fact that it's not just a description of what happens; you can deploy that flowchart to the process engine and it actually runs the process. It enables the automation of the business process with little or no programming. In addition to executing the process logic, it's generating the data that can be used to continuously monitor performance and KPIs.

BPM also addresses issues critical to IT and business drivers, including the following:

- Managing end-to-end, customer-facing processes
- Consolidating data and increasing visibility into and access to associated data and information
- Increasing the flexibility and functionality of current infrastructure and data
- Integrating with existing systems and leveraging emerging technology
- Establishing a common language for business-IT alignment
- Human task management
- Business rule management
- Forms, screen flows, and other user interfaces
- Modeling and simulation analysis

BPM provides a dialogue between the business and IT that can result in real collaboration. Process modeling is typically the responsibility of the business function, documenting and defining the process without implementation detail but with enough to define performance goals and to project performance metrics by the use of simulation analysis. Then, using the BPMS, the model automatically generates a skeleton design to be used by IT for the implementation. The process designer adds implementation detail to the skeleton created by the model and then deploys the design to the process engine, which runs the process, automates human workflow, integrates external applications using the middleware, and implements business rules. While the process is running, it is logging data that is analyzed by the BPMS's performance management component, computing many of the same KPIs that were projected initially in the model; displaying management dashboards of charts, alerts, and automated escalation procedures; and so on. In many BPM suites, the performance data can be fed back into the model for a new round of performance enhancement, simulations, and so on.

There are several reasons to invest in a BPMS:

1. It forces you to rethink your processes and add flexibility in applying changes. In today's business world, with the Internet, smartphones, wireless, Web services, globally distributed functions, and outsourced operations, you need to constantly rethink how your processes function and how you are keeping up-to-date and ahead of the competition.

2. You need to do more with less or the same. Higher efficiency, shorter cycle times, and higher volume with no increase in headcount or facilities—these are the realities of business today. The ability to apply process automation is key, and not just within specific functions, but for the entire value chain process.

3. It provides for compliance and control and the consistent use of documented processes across the organization.

4. It provides agility—the ability to deal with changes in the business environment, changing technology, bringing new products and services to market quickly, and responding to competitive pressures.

5. It enables a cycle of continuous improvement by collecting data and providing results to the process owners.

BPM AND BPMS AS A BEST PRACTICE

Studies by the American Productivity and Quality Center (APQC) found that "BPM is the way best-practice organizations conduct business." They also found that even though companies may be mature in their application of BPM, technology is vital to their continued success.

Chapter 4 of this book stated that the typical BPM cycle has four components—document, assess, improve, and manage. You can see that the application of a BPMS provides the mechanisms required for the automation of this entire cycle. Simply stated, the BPMS provides a vehicle for the successful implementation of BPM into an organization.

Here's a quick review. The document stage identifies the value chain processes and the process metrics. These process flows are the essential starting point for the BPMS. The assess phase monitors performance, which is embedded in the BPMS. The improve phase identifies processes or process segments needing improvement. The data is provided by the BPMS, and the ability to identify problem areas, make changes, and run simulations is made significantly easier through use of it. The manage phase requires that you continually monitor and maintain control of the value chain processes, which is likewise provided by the BPMS.

Figure 8-2 illustrates the Business Process Management life cycle phases and shows the overlap with what a BPMS provides. You can clearly see how a BPMS links directly to any effort to implement BPM in an organization by providing all of the necessary elements and automating them for ease of use.

The BPMS platform automates processes, moves information, and includes people in the process when skills are required. In this software environment, processes can be designed or redesigned, simulated (as discussed previously), executed, monitored, and analyzed, allowing organizations to take the processes they've designed using any of the methodologies discussed and implement them in the most efficient way possible. The BPMS coordinates business processes between people and IT systems, putting the data in context and converting it into meaningful business information instead of simply routing it automatically from system to system. It delivers useful information to participants in the process, enabling them to make informed decisions in a timely manner. From a management viewpoint, the BPMS collects and analyzes process metrics—for example, cycle time

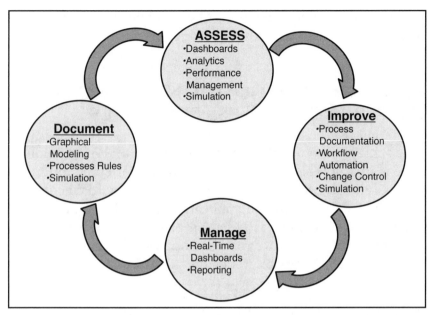

FIGURE 8-2.   Business Process Management life cycle phases and the BPM suite.

and cost—so that performance and value can be tracked with accuracy.

The BPMS platform monitors business activity across multiple software systems. When changes occur in normal operations, this platform determines whether the event requires people or system actions and either launches automated processes or prompts employees to respond, guiding them through their participation in the process. The BPMS also provides executive dashboards that supply real-time reports on process performance, how those processes affect the business, and where the processes can be further improved.

Process-focused initiatives have been around for years—for example, Business Process Reengineering (BPR), Lean approaches, and Six Sigma. Continuous process improvement is critical to long-term business success, but these programs have not delivered to their full potential—not because they were necessarily misguided methodologies but because they had no direct link to bottom-line profitability nor any direct connection to technology to support the ideals of continuous process improvement across the corporation.

BPM AND BPMS: COMMON GOALS FOR IMPROVING BUSINESS PROCESSES

The combination of BPM methodology and a BPM technology suite is a natural pairing for making process-oriented initiatives successful. Table 8-1 gives a summary of the way BPM concepts are enhanced through the application of a BPM suite. While the BPM methodology is based on proven business principles that integrate corporate goals into process execution efforts, BPM software provides a flexible technology as the sound

TABLE 8-1. BPM and BPM Suite: Common Goals for Improving Business Processes

	BPM	BPM SUITE
Approach	Analytical strategy for generating value through core business processes	Automation and optimization environment enabling business processes
Data	Identifies key process indicators and uses them to identify improvement opportunities	Accesses data from enterprise systems for monitoring and improvement
Process improvement	Process improvement opportunities gained through root cause analysis tools	Process flows graphically defined, showing people-system interactions
Process improvement implementation	Documented system for implementing changes and measurements	Improved automation and faster execution of documented changes
Measurement	Key metrics routinely monitored	Dashboards of KPIs generated

foundation for designing and executing these ideas throughout the enterprise.

Consider the common goals of BPM and BPMS to be reduced costs, increased profits, and greater customer satisfaction through business process improvement. When you consider the challenges facing today's organizations—data housed in disparate transaction-based systems throughout the organization and manual processes that lack standardization and cannot easily be monitored—the analytical approach of BPM, combined with the automation and optimization approach of a BPMS, provides the solution to an organization's process and technology challenges, while tying the solution directly to the corporation's financial goals. The following sections discuss some specific ways that a BPM suite complements BPM to solve process and data challenges, while delivering a return on investment.

DATA MONITORING

The typical company must manage its core processes and, realistically, numerous subprocesses, plus all the data required to run its operations. It has spent millions of dollars acquiring enterprise-wide systems aimed at planning, budgeting, forecasting, managing, and analyzing its business, and it has spent additional time and money training its employees to use these systems. The challenge is that many processes and correlating data span multiple transactional systems, making metrics gathering and analyzing cumbersome, time consuming, and sometimes even impossible, despite all of the previous investments in technology. Typical BPM suites monitor business process data across the systems related to that process. They look for business events that fall inside or out of defined business rules and launch warnings in response to those events. A BPMS has integration components that allow it to integrate with other applications, pushing and pulling data between any application and itself. These two features allow a company to extend its process improvement activities across the enterprise, despite the location of the process or the associated data and people.

PROCESS IMPROVEMENT DESIGN AND EXECUTION

One of the challenges of improving processes is that they often span multiple IT systems. As a result, improving a process requires making changes to multiple applications that the process spans. Changing large applications can be cost-prohibitive—requiring money and time from key resources that have already been spent on initial implementations.

To meet this challenge, people tend to get heavily involved in the process, with inconsistent and inefficient manual work-around procedures. The opportunity for errors and excessive process cycle time adds up to waste. A BPMS platform solves this problem by designing and executing processes that reach across multiple applications. It allows the process to bridge applications, gathering data as needed, regardless of the source of the data. It monitors the process for any event that falls outside of its normal parameters. When an event occurs, it uses defined business rules to either launch an automated subprocess or bring a person into the process by sending a message that action is required.

MEASUREMENT: DASHBOARDS

The BPMS collects data and puts it into the context of a business process. It must be measured and analyzed so that it becomes meaningful information that process owners can use to make informed decisions about the process performance. In addition, new data may need to be created about the performance of the process itself so that executives can determine where processes are lagging and where they have been successfully improved. This is critical for continuous improvement because it gives specific information about process performance to the process owners and executives in an enterprise, providing them with the information they need to strategically respond to changing business conditions. This data can then be displayed as a dashboard for a complete picture of business processes on a regular basis, giving the process owners and executives insight and control that was previously impossible. This capability gives executives the metrics they need to make decisions about ongoing process improvements.

CONCLUSION

BPM's analytical approach to continuously improving processes across the corporation can be enhanced even further by using BPM software to design and execute the core business processes. The BPMS monitors systems for business events and pulls relevant data from various systems, presenting the information to the appropriate person at the right time, so that processes are executed quickly and consistently, and errors are greatly reduced. It provides a graphical authoring environment so that core business processes can be quickly and easily defined or modified and deployed by nontechnical business process owners. The BPMS gathers metrics as process events happen, delivering to executives the information they need to analyze, modify, improve, and control processes in real time.

Business Process Management software has changed the equation. A BPMS is designed to enable rapid builds of new capabilities and delivery of value much faster than previous IT deployments. Modern BPMS platforms provide rich functionality for modeling, analysis, simulation, execution, control, and management of processes. A BPMS enables complex process interactions and connects internal and external applications so that processes can flow seamlessly across business lines and functions. A BPMS provides for sophisticated process management, inclusive of existing systems and applications.

CHAPTER 9

BPM RESOURCES

There are many resources available to those interested in pursuing the implementation of a BPM system. These resources include academia, advisory services, conferences, consulting service providers, professional societies, publications, software vendors, and training providers.

THE BPM COMMUNITY

Academia—Research and Education

- Babson College, Process Management Research Center
- Boston University
- Illinois Institute of Technology
- Indiana University
- Northwestern University
- Massachusetts Institute of Technology
- Stevens Institute of Technology
- University of Chicago
- Widener University

Advisory Services

- Aberdeen Group
- BP Trends
- BPMInstitute.org
- Delphi Group

- Forrester
- Gartner

Conferences
- BPMInstitute.org's conference
- Gartner Business Process Management Summit
- IIR BPM Conference
- Software vendor user conferences
- IQPC BPM and Six Sigma conferences
- ASQ Six Sigma Conference

Consulting Service Providers
- Accenture
- BearingPoint
- Cap Gemini Ernst & Young (CGEY)
- Computer Science Corporation (CSC)
- Deloitte
- IBM
- KPMG
- Many boutique consultancies
- Many software vendor consulting services

Professional Societies
- Association of Business Process Management Professionals (ABPMP)
- American Production and Inventory Control Society (APICS)
- American Society for Quality (ASQ)
- American Society of Training Directors (ASTD)
- International Institute of Business Analysts (IIBA)

Publications

- Newsletters, magazines, online resources
- BP Trends publications
- BPMInstitute.org
- BrainStormCentral.org
- *CIO* magazine
- *Intelligent Enterprise* magazine
- iSixsigma.com
- Process Excellence Network publications

Training Vendors

- BP Trends
- BPMInstitute.org
- Performance Design Labs
- Rummler-Brache Group

PROCESS IMPROVEMENT TOOL KIT

Here are just a few of the key tools that can be used while executing a process improvement project. You can find further information about these tools, including diagrams and instructions, at the iSix Sigma Web site (http://www.isixsigma.com).

TOOLS FOR GENERATING IDEAS AND ORGANIZING INFORMATION

- *Brainstorming.* This is a method for generating ideas. Ground rules such as "no idea is a bad idea" are typical. The benefit of brainstorming is the power of the group in building ideas off of each other's ideas.
  - o It serves as a starting point.
  - o Its basic purpose is to come up with a list of options for a task or solution.

o   The list will be shortened into a final selection.

o   Brainstorming can be used again to list possible measures and, still later, to come up with creative improvement solutions.

- *Affinity diagramming.* This tool is used to organize large amounts of data into logical categories based on perceived relationships and a conceptual framework. It is often done using "sticky notes" in a brainstorming exercise, and then the data is grouped by a facilitator and workers. The final diagram shows the relationship between the issue and the category. Then categories are ranked, and duplicate issues are combined to make a simpler overview.

  o   This is a common follow-up to brainstorming that helps to evaluate ideas.

  o   The best method is for people to remain quiet, grouping ideas without speaking.

- *Multivoting.* This tool is used to arrange and rank by importance a list of ideas, problems, common causes, and the like. The list typically consists of a few (usually three to five) controllable items. It is a group effort in which every member of the group is allowed to assign a number ranking the importance of each item. Those items receiving the highest rankings from the group should receive consideration first.

  o   This method is used to narrow down a list of ideas or options.

  o   It serves as a follow-up to brainstorming.

- *Tree diagram.* This breaks down ideas into progressively greater detail. It shows the links between or the hierarchy of the ideas that have been brainstormed. The objective is to partition a big idea or a problem into its smaller components, making the idea easier to understand or the problem easier to solve. This method might also be used to tie major customer needs to more specific requirements.

- *High-level process map (SIPOC diagram).* SIPOC stands for "suppliers, inputs, processes, outputs, and customers." You obtain inputs from suppliers, add value through your process, and provide an output that meets or exceeds your customer's requirements. It's a method that keeps you from forgetting something when you are mapping processes.

  o   This is the preferred method for diagramming major processes and identifying possible measures.

  o   It shows major activities, or subprocesses, in an organizational process, along with the framework of the process.

  o   It helps to define the boundaries and critical elements of a process without getting into so much detail that the big picture is lost.

- *Flowchart.* A flowchart is a graphical representation of a process from start to finish, showing inputs, pathways and circuits, actions or decision points, and, ultimately, completion. It can serve as an instruction manual or a tool for facilitating detailed analysis and optimization of workflow and service delivery.

  o   It shows the details of a process including tasks, procedures, alternative paths, decision points, and rework loops.

  o   It shows how a process currently works or how it should work.

- *Cause-and-effect diagram (fishbone diagram).* This tool is used to solve quality problems by brainstorming causes and logically organizing them by branches.

  o   It is used to brainstorm the causes of problems.

  o   Causes that lead to other causes are linked in a structure tree.

  o   They do not tell the root cause. They help you develop an educated guess about where to focus measurement and further root cause analysis.

Tools for Gathering Data

- *Sampling.* This offers a much more realistic approach than counting every measure that goes into a process. Counting or measuring everything can be difficult and expensive.

- *Operational definitions.* An operational definition can be defined as a clear and understandable description of what is to be observed and measured, so that different people collecting, using, and interpreting data will do so consistently. An operational definition is a concept to guide what properties will be measured and how they will be measured.

- *Checksheets and spreadsheets.* These are used to collect and organize data. They ensure that the appropriate data is captured and that all necessary facts are included, such as when an event happened, how many times, and what customer was involved. These can vary from simple tables and surveys to diagrams used to indicate where errors or damage occurred.

- *Measurement systems analysis (MSA).* This is an experimental and mathematical method of determining how much the variation within the measurement process contributes to overall process variability. It helps measure the effectiveness of gauges, rulers, and other measurements.

- *Voice of the customer (VOC) methods.* "Voice of the customer" is the term used to describe the stated and unstated needs or requirements of your customers. The voice of the customer can be captured in a variety of ways: direct discussion or interviews, surveys, focus groups, customer specifications, observation, warranty data, field reports, and so on.

  o These methods collect external customer input, assess and prioritize requirements, and provide ongoing feedback to the organization.

  o Tools may be simple or sophisticated market research methods, requirement analysis concepts, and newer technologies, such as data warehouses and data mining.

TOOLS FOR PROCESS AND DATA ANALYSIS

- *Process-flow analysis.* This method uses a map or flowchart of a key work process to scrutinize the process for redundancies, unclear handoffs, unnecessary decision points, delays, bottlenecks, defects, and rework. It can be one of the quickest ways to find clues about the root causes of problems.

- *Value- and non-value-added analysis.* Each step in a detailed process map is assessed on the basis of its real value to external customers. It's never possible to eliminate all non-value-adding activities because some are in place to protect the organization or to meet legal requirements. This process helps remove things that are unnecessary in a process and are a drain on resources.

- *Charts and graphs.* These constitute a visual display of the data. Charts and graphs of various types offer a different picture of the data. These charts can include:

  - *Pareto chart.* This is a specialized bar chart that breaks down a group by categories and compares them from largest to smallest. It is used to determine the biggest pieces of a problem or contributors to a cause and which problems have the biggest impact. The Pareto chart utilizes the 80–20 rule, which states that 80 percent of the results come from 20 percent of the causes.

  - *Histogram.* This shows the distribution or variation of data over a range of size, age, cost, length of time, weight, and so on. In analyzing a histogram, you look for the shape of the bars or the curve, the width of the spread or range, or the number of modes.

  - *Run chart.* This shows how trends are changing over time.

  - *Scatter diagram.* This looks for direct relationships between two factors in a process to see if there is any correlation. If two measures show a relationship, then one might be causing the other. When an increase in one factor matches an increase in the other, it is called a *positive correlation*. When an increase in one causes a decrease in the other, this relationship is called a *negative correlation*.

TOOLS FOR IMPLEMENTATION AND PROCESS MANAGEMENT

- *Project management methods.* Well-run organizations recognize early on the importance of strong project management skills: planning, budgeting, scheduling, communication, people management, and technical project management tools.

- *Potential problem analysis and failure mode and effects analysis.* These are two of the methods that are applied both in implementing new processes and in running them every day. Both start with brainstorming the many things that could go wrong. Then the potential problems are prioritized. Finally, the biggest risks are mitigated by looking for ways to prevent them from happening, as well as ways to limit contingencies.

- *Stakeholders analysis.* This involves identifying the people and groups that need to be considered, their likely views on the project or solution, and approaches to gaining their input and/or support.

- *Force field diagram.* This shows the relationship between factors that help promote a change and those that oppose or create resistance to it. It is used to develop plans to build support for critical change.

- *Process documentation.* Effective, clear, not overly complex process documentation, such as process maps, task instructions, measures, and so on, are created.

- *Balanced scorecards and process dashboards.* These provide a summary of critical measures that give real-time feedback and promote prompt attention to issues and opportunities.

These tools typically feature both output (Y) and process and input (X) measures and go beyond traditional financial data. Process improvement activities have placed new attention on the ability of people throughout an organization to keep tabs on current performance, trends, and issues of key indicators in a process.

TOOLS FOR STATISTICAL ANALYSIS

- *Tests for statistical analysis.* This method looks for differences in groups of data to see whether they are meaningful. These tests include t-tests, chi-squares, and analysis of variance (ANOVA).

- *Correlation and regression.* These tools test for the presence, strength, and nature of the links among variables in a process or a product. They include regression coefficients, simple linear regression, multiple regression, and surface response tests.

- *Design of experiments (DOE).* DOE is a collection of methods for developing and conducting controlled assessments of how a process or a product performs, usually testing two or more characteristics under different conditions. In addition to helping target causes of a problem, DOE can be essential to get maximum benefit out of solutions, called *optimizing results.*